PRAISE FOR
YOUR BODY IN
BALANCE

"In YOUR BODY IN BALANCE, Dr. Neal Barnard brings the latest nutritional science down to bite-sized truths that will quickly transform nearly any life plagued by hormone-dependent illness into one of vibrant wellness. Whether it's cramps, cancer, infertility, impotence, moodiness, or menopause, Dr. Barnard masterfully navigates a clear path to your healthiest self."

—KRISTI FUNK, MD, FACS, breast cancer surgeon, co-founder of Pink Lotus Breast Center, and bestselling author of *Breasts: The Owner's Manual*

"Dr. Neal Barnard is one of the most important authorities of our time on nutrition, diet, and health, and YOUR BODY IN BALANCE is the book that can and will finally change your health for good. Dr. Barnard walks us through the most common and troublesome ailments which so many of us struggle with—hormone fluctuations, thyroid conditions, chronic and terminal illness, and mood disorders—with an eye toward research and solutions based in foods commonly available to us. With recipes included, this may just be the only book you need on your shelf in order to change your body, your health, and your life."

—MAYIM BIALIK, PhD, neuroscientist and actor

"Finally, an intelligent guidebook that speaks up about the potent role our hormones play in every aspect of our lives. As an Olympic athlete, I never really thought of this connection until I dropped all animal foods from my diet and turned to whole plant foods to fuel my training. Almost overnight, my adaptation to training sped up (which is a function of our immune system directed by hormones), my PMS subsided, my breakouts cleared up, and my ability to focus and calm my nerves greatly improved. Dr. Barnard takes the reader on a revealing journey to uncover what's really behind some of our most common physical ailments and how to fix them in the blink of an eye."

—DOTSIE BAUSCH, Olympic silver medalist, eight-time U.S. National Champion, former world record holder, two-time Pan American Gold, and founder of Switch4Good

"YOUR BODY IN BALANCE is an incredible resource. If you have ever wondered how food impacts your fertility, erectile function, thyroid function, skin, hair, and so much more, wonder no longer. This book shows us how what we eat can literally change our hormones, our mood, and our health! Happy eating!"

—ROBERT OSTFELD, MD, MSc, FACC, director of preventive cardiology at Montefiore Health System in New York

YOUR BODY IN BALANCE

YOUR BODY IN BALANCE

The New Science of Food, Hormones, and Health

Neal Barnard, MD, FACC

With Menus and Recipes by Lindsay S. Nixon

First published in the United States by Grand Central Publishing, Hachette Book Group in 2020

First published in Great Britain by Sheldon Press in 2020, an imprint of John Murray Press, a division of Hodder & Stoughton Ltd,
An Hachette UK company

1

The information and program herein is not intended to be a substitute for medical advice. You are advised to consult with your health care professional with regard to matters relating to your health, and in particular regarding matters that may require diagnosis or medical attention.

A CIP catalogue record for this title is available from the British Library

Paperback ISBN 978 1 529 34443 1
eBook ISBN 978 1 529 34444 8

Printed and bound in Great Britain by Clays Ltd, Elcograf S.p.A.

John Murray Press policy is to use papers that are natural, renewable and recyclable products and made from wood grown in sustainable forests. The logging and manufacturing processes are expected to conform to the environmental regulations of the country of origin.

John Murray Press
Carmelite House
50 Victoria Embankment
London EC4Y 0DZ

www.sheldonpress.co.uk

Contents

Part III FEELING BETTER AGAIN

A Note to the Reader

This book will introduce you to the power of foods to protect and restore your health. It's easy to put that power to work, and the payoff is huge, as you will soon see! Even so, let me mention two important points:

• **If you have a health condition or use medications, see your health care provider.** Often, people need less medication when they improve their diets. This is common for diabetes or high blood pressure, for example. Sometimes, they can discontinue their drugs altogether. But do not change your medications on your own. Work with your clinician to reduce or discontinue your medicines if and when the time is right.

• **Get complete nutrition.** Plant-based foods are the most nutritious foods there are. Even so, you will want to ensure that you get complete nutrition. To do that, include a variety of vegetables, fruits, whole grains, and legumes in your routine, and I would suggest a special focus on green leafy vegetables. And be sure to have a reliable source of vitamin B_{12} daily, such as a simple B_{12} supplement, fortified cereals, or fortified soymilk. Vitamin B_{12} is essential for healthy nerves and healthy blood. You will find more details in Chapter 12, "A Healthy Diet."

UK and US Measures:
Conversion Charts

Oven Temperatures

Gas Mark	Celsius	Fahrenheit
1	140	275
2	150	300
3	165	325
4	177	350
5	190	375
6	200	400
7	220	425
8	230	450
9	245	475
10	260	500

Volume

Metric	Imperial	Us Cups
250 ml	8 fl oz	1 cup
180 ml	6 fl oz	3/4 cup
150 ml	5 fl oz	2/3 cup
120 ml	4 fl oz	1/2 cup
75 ml	2 1/2 fl oz	1/3 cup
60 ml	2 fl oz	1/4 cup
30 ml	1 fl oz	1/8 cup
15 ml	1/2 oz	1 tbsp

Weight

Metric	Imperial
15g	1/2 oz
30g	1 oz
60g	2 oz
90g	3 oz
110g	4 oz
140g	5 oz
170g	6 oz
200g	7 oz
225g	8 oz
255g	9 oz
280g	10 oz
310g	11 oz
340g	12 oz
370g	13 oz
400g	14 oz
425g	15 oz
450g	1 lb

Introduction

Some of the most troubling health problems have a surprising solution. Weight problems, infertility, menstrual cramps, diabetes, thyroid problems, acne, hot flushes, and many others are related to the foods you eat. In each case, whether you know it or not, foods are changing your hormones, making you feel great or terrible, vigorous or lethargic, pain-free or miserable.

The key is this: You can control your hormones—and the problems they cause—by the food choices you make every day. It is surprisingly easy, and I will show you how to do it.

Like many women, Robin had menstrual cramps that were off the scale for a day or two every month, until she learned of a straightforward dietary shift that eased her pain and allowed her to function again.

Elsa had given up on raising a family, until she found that her fertility had been affected by what she had been eating. After tuning up their diet, she and her husband now have three children.

Ray had assumed that erectile dysfunction was his new normal, until an easy diet change gave him back his sex life.

Nancy's thyroid gland was underperforming, keeping her overweight and fatigued, until she learned a new way of getting back in balance—all based on a menu change.

Kim was chronically down in the dumps with a serious and unremitting depression, but was finally able to lift the black cloud hanging over her when she learned the connection between food and mood.

Twin sisters Nina and Randa were embarrassed about their severe acne, until a nutritional adjustment tackled it nearly overnight.

Nutrition was the surprising answer to Tony's prostate cancer, Katherine's endometriosis, Mike's thyroid problems, Bob's diabetes, and Mary-Ann's menstrual misery.

All of these people have something in common: Their hormonal systems had been chronically malfunctioning. Yours may be, too.

So let me encourage you to run—do not walk—to a healthier way of eating. I'll show you how foods cause hormones to go haywire and how the right foods can fix that.

Hormones? Really?

Your body's hormones direct how your organs work. Like a conductor directing a symphony orchestra—quicken the tempo or slow it down, more violin, less bass—your hormones turn your metabolism up or down, alter your moods, control your reproductive function, and affect how you store body fat and how you burn it.

The people I described above all suffered from hormone haywire. They had no idea what was wrong, much less how to fix it.

Their answer was in foods. Yours may be, too. Your hormonal systems respond to changes in your diet, and the range of health problems that are caused by hormones—and that can be successfully treated by using foods in a smarter way—is truly surprising. Equally surprising is *how fast* things can improve with a menu adjustment: Cramps gone in the very first month after a diet change. Fertility restored within a few months. Diabetes starting to improve in a matter of days. While everyone's experience is different, these are all things we have seen routinely.

Estrogens and testosterone—the female and male hormones—can be ratcheted up or down based on the foods you eat. In turn, that affects fertility for both men and women, as well as a woman's monthly cycle, menopausal symptoms, and the risk of certain cancers.

Insulin is the hormone that is off-kilter in diabetes, and it also struggles to do its job in women with polycystic ovary syndrome. With the right foods, insulin works much better.

Research also shows that foods can influence thyroid function—which affects weight, mood, the resilience of your skin and hair, and many other aspects of health. We are now coming to understand how to change the diet to get the thyroid gland in balance.

If you thought that erectile dysfunction was caused by performance anxiety, I will show you the real cause and the easy, healthful solution that has nothing to do with pills or gadgets and all to do with what you had for breakfast.

Some of the most serious conditions we face—weight problems, diabetes, and cancer of the breast, ovary, uterus, or prostate—are related to foods that are causing hormone haywire. And for less serious conditions—the health of your skin and whether you keep a full head of hair—hormones play a key role, too. Foods turn the dials on virtually every biological function.

Why Didn't My Doctor Tell Me?

If you were hoping to learn about these effects from your personal physician, you may end up frustrated. Much of the information we will present in this book is a new scientific frontier. It is possible that your doctor may not have heard of it or may not realize the extent to which problems can be solved with nutritional adjustments.

In other areas, solid research was published long ago but has been ignored by most doctors. Let's face it: A focus on nutrition takes doctors out of their comfort zone. Part of the blame goes to the chronic lack of nutrition education in medical schools. Another part goes to the fact that most continuing medical education programs are sponsored by drug companies, which have pushed to the sidelines anything that is not sold at the pharmacy counter. This is not a problem just for cramps and fertility, but also for diabetes, heart disease, cancer, migraine, and inflammatory conditions like rheumatoid arthritis, where a diet change can be powerful, even lifesaving.

Sometimes doctors are unduly skeptical about whether their patients are really interested in improving their eating habits—even when their patients tell them loudly and clearly that, yes, they would much rather improve their diets than take drugs for the rest of their lives. Many doctors are much more comfortable with a prescription pad than with a knife and fork.

At the Physicians Committee for Responsible Medicine, a nonprofit organization that I founded in 1985, we are working to change that. Our research is teasing apart the relationships between foods and health and giving doctors and patients new and powerful tools. Our continuing medical education courses, conferences, mobile apps, and books give doctors the nutrition knowledge they need. More and more health professionals are using this knowledge to empower their patients.

It should be said that all the health issues covered in this book are still under active exploration. There are many areas where our scientific journey is far from over. I will share with you what we know now and will also give you a glimpse into the new possibilities that are being tested.

How to Use This Book

Looking at the table of contents, you may be tempted to go straight to the chapter that describes your most pressing health issue. Feel free to do that. But I would encourage you also to read this book from cover to cover—including sections on conditions that may not be yours at the moment—because, as you will see, certain themes emerge again and again. The same nutritional solutions that help cramps or infertility can also have a powerful influence on thyroid problems or even cancer.

My hope is that, with guidance from this book, you will achieve a whole new level of vitality. I also hope that you will pass this information along to family members, friends, and others who can benefit from it.

I wish you the very best of health.

YOUR BODY IN BALANCE

Part I

SEX HORMONES: FINDING YOUR BALANCE

CHAPTER 1

Foods for Fertility

I have good news and bad news," Elsa said. She was going to leave our research study, and I could tell she wasn't the slightest bit sad about it.

When she volunteered for our research, she did not imagine that it would change her life forever. The study was a test of how foods could reduce menstrual cramps. It followed up on my observations that a menu tune-up sometimes makes cramps go away. (More about that in Chapter 2, "Curing Cramps and Premenstrual Syndrome.") Elsa had suffered with menstrual pain for years and was eager to see how a diet change might help.

Like all the volunteers, she was asked not to use birth control pills during the study, because birth control pills are hormones that could obscure the effects of the diet change. That was not an issue for her, she told us. She didn't actually need birth control at all; she and her husband had long since given up the idea of having a baby. They had been evaluated medically, and the problem was hers. She was infertile.

Joining the study, she followed our advice on how to choose foods, which you will read about shortly. It was not especially complicated. And soon, something unexpected happened.

"Good news and bad news?" I asked. "Tell me! What is it?"

"Well, the bad news is that I have to drop out of the study," she said. "I'm sorry about that, but the good news is that *I am pregnant!*"

Yes, her so-called infertility was over. She was indeed pregnant and soon gave birth to a beautiful, healthy baby. She never imagined, after all the medical tests, monitoring, and treatments she and her husband had been through, that a simple diet change would set things right.

Years later, I was giving a lecture in the Midwest, and I was surprised to see Elsa in the audience. She had heard that I was coming to town and wanted to let me know how she was doing and to introduce me to her *three* children.

Katherine

Katherine grew up in Louisiana. As an aerospace engineer for the Air Force, she was one of the first to be deployed to Iraq in 2003.

In the war, government rations were limited, to say the least. And when Katherine's tour of duty ended and she went back home to Louisiana, her friends were eager to reunite her with her old favorite foods: cheeseburgers, macaroni and cheese, shrimp, and gumbo.

Not surprisingly, she gained weight. And as time went on, she experienced something else—pain in her abdomen. The pain gradually worsened and came in waves, waxing and waning with her menstrual cycle, returning month after month.

She saw her doctor to try to find out what was wrong. After several tests, her gynecologist recommended a laparoscopy. He made a small incision in her abdomen, inserted a scope, and looked around. And he spotted the cause of her pain. She had *endometriosis*, a condition in which cells that normally line the uterus have migrated to other parts of the abdomen, where they implant and cause pain that can be excruciating. Because the out-of-place cells become inflamed and can disrupt the anatomy of the ovaries and fallopian tubes, endometriosis can also lead to infertility.

Finding the cause of her pain was one thing. Curing it was another. Typical treatments include painkillers, hormone treatments, and laparoscopic surgery. They may or may not work. When all else fails, doctors can perform a hysterectomy.

Things were not going well for Katherine. The pain worsened over time to the point where she was essentially out of commission for a day or two every month. None of the treatments helped. Eventually she had to make a decision. Although she and her husband had hoped to raise a family, she could not live like this, and she scheduled a hysterectomy.

But before she had the surgery, a friend made a helpful suggestion. Why not try a diet change? After all, foods can influence the body's hormones, which is why many women with breast cancer change their diets. Maybe it would work for endometriosis, too. So, she made an appointment for some nutrition guidance and revamped her diet. Following her nutritionist's advice, she eliminated animal products and explored the world of vegetables, grains, beans, and fruits and all the meals they turn into.

Her new foods turned out to be tasty. But more importantly, she started feeling better. She lost weight, her energy improved, and the pain started to loosen its grip. With each passing week, she felt more like her old self.

Six weeks later, her doctor repeated the laparoscopy. Again, he made a small incision in her abdomen, inserted the scope, and carefully looked around. He then sewed up the incision, sent Katherine into the recovery room, and walked out to the waiting room to find her husband. The doctor explained that, to his surprise, her endometriosis was practically gone. So much so, in fact, she no longer needed a hysterectomy at all.

Her husband was not so surprised. He told the doctor about how Katherine had revamped her diet and how it had helped her enormously. She had been feeling better and better.

The doctor was having none of this. Foods do not cause endometriosis, he explained. And there is no way that a diet change could make it go away. There was only one possible explanation, the doctor said. This was *a miracle!*

Katherine laughs about that today. But it did feel pretty miraculous. She lost fifty-five pounds in six months, her pain went away completely, and her emotional ups and downs smoothed out, too. Most miraculous of all, she never had the hysterectomy. She and her husband now have three children.

Foods and Fertility

This may not be a time of life when fertility is an issue for you. And these days, many people are opting not to have children anyway, because of time, finances, concerns about overpopulation, or other reasons, and the days when people were pressured to have children have mostly gone. Nonetheless, let me encourage you to read this chapter, as its lessons extend to many other conditions as well.

In theory, fertility is simple. An egg from Mom meets a sperm from Dad, and they cozy up in a place where they can unite and grow. But things can go wrong, both for Mom and for Dad.

For about 10 to 15 percent of couples trying to conceive, it just does not happen after more than a year of trying. Many spend a fortune on medical evaluations and treatments of all kinds. Beyond the financial cost is the burden of medical visits, examinations, treatments, and the risk that romance will turn into a sterile medical exercise.

The most common reason for infertility in women is a lack of ovulation. The ovaries are stubbornly refusing to release an egg.

For men, abnormalities of the testes, the hormones that control them, or the ducts that carry sperm along can all limit fertility.

There can also be problems with bringing the egg and sperm together: At the bottom of the uterus, the cervix admits sperm while, at the top, the fallopian tubes are where the egg is fertilized. Sometimes these structures do not form properly. Other times, infections interfere. Or perhaps trauma or other abdominal problems intervene. Yet foods have surprising effects on fertility for both men and women. And if food choices are causing the problem, the solution can be straightforward. Let's start with ovulation.

Foods and Ovulation

The ovary is an amazing structure. One of the most amazing things about it is that, by the time you were born, your ovaries were already developed, with millions of eggs just about ready to go. That's right. While you were still a baby in your mother's womb, your own tiny ovaries had already developed the eggs that could become *your* babies. That means that things your mother was exposed to during pregnancy could easily have affected you, your ovaries, and your eggs.

During your reproductive years, your ovaries are ready to release an egg each month. But that process is easily derailed. If your hormones are out of balance, ovulation may not occur at all.

Millions of Hormone Factories

Pinch a bit of flesh on your thigh. If you could reach into your thigh and pull out one single fat cell and examine it, what would it look like? We tend to think of fat cells as lifeless little bags of stored calories. The truth is very different. Fat cells are busy hormone factories. Every minute of every day, they crank out hormones.

Starting with hormones produced by your ovaries and adrenal glands, fat cells turn these raw materials into the male hormones and female hormones that control your basic biology.*

The more body fat you have, the more hormones they produce. That can get your hormone balance out of kilter and interfere with your fertility.

Having extra body fat will not help your fertility one bit. In Harvard's Nurses' Health Study II, researchers examined fertility in a large group of women, based on their body weight.[1]

Fertility was greatest for those who were a bit on the thin side of the normal weight range. At even slightly higher weights— even for women within what we would consider a healthy body weight—infertility was more common. For seriously overweight women, fertility problems were nearly three times more common.†

The reason, we believe, is that fat cells cause hormones to go haywire. They actually do this in two ways:

First, as we have seen, fat cells produce extra female sex

---

* Your adrenal glands, above your kidneys, produce a compound called *androstenedione*. Fat cells convert this compound to testosterone—the "male hormone"—and then to estradiol, a common form of estrogen. It can also be converted to estrone, another form of estrogen.

The word *estrogen* refers to a group of hormones that influence a woman's monthly cycle and reproductive function. In young women, the main estrogen in the bloodstream is called *estradiol*. After menopause, the predominant estrogen in a woman's bloodstream is *estrone*.

† In the Harvard study, body weights were recorded as *body mass index*, or BMI, which is a way of adjusting for how tall you are. A healthy BMI is between 18.5 and 25 kg/m². For a woman who is five feet five inches tall, this corresponds to a weight range from 111 to 150 pounds.

Fertility was greatest for those with a BMI between 19 and 22 kg/m² (that would be 114 to 132 pounds for a woman who is five feet five inches). For women with BMIs over 30 kg/m² (180 pounds at five feet five inches), fertility problems occurred three times more frequently.

You can check your own BMI using the Physicians Committee for Responsible Medicine's BMI calculator. Just visit www.pcrm.org/weightloss, plug in your height and weight, and you will see your BMI.

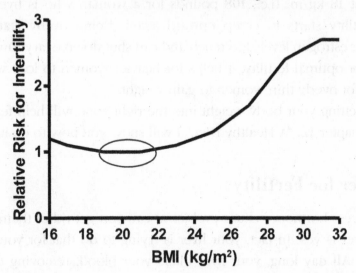

Risk of Infertility

Rich-Edwards JW, Goldman MB, Willett WC, et al. Adolescent body mass index and infertility caused by ovulatory disorder. *Am J Obstet Gynecol.* 1994;171(1):171–177.

hormones (estrogens) and also extra male sex hormones (androgens) and send them into a woman's bloodstream.

Second, body fat reduces the amount of *sex hormone–binding globulin,* or SHBG, in the bloodstream. SHBG is a very helpful protein molecule. It circulates in your bloodstream and holds on to sex hormones, keeping them inactive until they are needed. Think of SHBG as a fleet of microscopic aircraft carriers and estrogens as tiny fighter planes. So long as the "planes" stay on the "carrier," they remain inactive. If you have plenty of SHBG in your bloodstream, your sex hormones will not be overactive, and that is good.

So if your fat cells are producing extra hormones and, at the same time, are reducing the amount of SHBG that would rein those hormones in, you'll end up with too much hormonal activity. So carrying extra weight can reduce fertility.

That said, thinner is better only up to a point. Below a BMI of about 18 kg/m² (i.e., 108 pounds for a woman who is five-five), infertility starts to creep upward again. Being overly lean can lower estrogen levels too much and can shut down ovary function. So for optimal fertility, it helps for heavier women to lose weight and for overly thin women to gain weight.

Getting your body weight into the right zone will help fertility. In Chapter 12, "A Healthy Diet," I will show you how to do it.

Fiber for Fertility

If excess hormones cause problems, can you get rid of them? The answer is yes. In fact, your liver is trying to do that for you right now. All day long, your liver filters your blood, removing things that do not belong there: toxins, medications, and other things, including hormones. Your liver sends them into a small tube, called the *bile duct*, which leads to the intestinal tract. There, fiber soaks them up and carries them out with the wastes. You are literally flushing away unwanted hormones.

There is plenty of fiber in beans, vegetables, fruits, and whole grains, and this healthy fiber escorts unwanted hormones out of your body. But animal products do not have fiber. If chicken, fish, dairy products, and eggs are a prominent part of your diet, you may not have enough fiber to soak up and remove unwanted hormones. That means hormones end up being reabsorbed back into your bloodstream. This process is called *enterohepatic circulation* (*entero* refers to the digestive tract, and *hepatic* refers to the liver). Hormones are ejected from the liver into the intestine, but then pass into the bloodstream and eventually back to the liver, where the whole process repeats, several times a day. Instead of flushing away unwanted hormones, your body keeps "recycling" them, leaving you with more hormones than you need.

What is the answer? Chuck out the animal products and bring in the vegetables, fruits, beans, and whole grains. Their fiber will capture the hormones in the digestive tract and carry them away once and for all. Instead of recycling unwanted hormones in your bloodstream, you will send them down the toilet. You will still have enough hormones for normal functioning, and getting rid of the excess is a big help.

Fiber does the same for cholesterol, as you may have heard from television commercials for breakfast cereals made from oats. The oat fiber carries cholesterol out with the wastes, gently lowering your blood cholesterol levels.

Boosting Fiber

The average American gets about 16 grams of fiber per day, according to U.S. government statistics.[2] For good health, we should double that figure, at the very least. Beans, vegetables, fruits, and whole grains can easily get you to 40 grams per day, a healthful goal.

Would this really work? The answer came from studies at Tufts University, UCLA, and the American Health Foundation. Women volunteers changed the fat and fiber content of their meals under controlled conditions and, yes, high-fiber, low-fat diets trimmed estradiol levels by anywhere from 10 to 25 percent. Estrone was tamed to about the same degree, and testosterone was reduced, too.[3,4,5]

Fiber is not all there is to it, though. Researchers have long taken an interest in the "Mediterranean diet" of southern Italy, emphasizing fruits and vegetables and favoring olive oil when oils were to be used, while de-emphasizing meats and dairy products. In 2006, Sicilian researchers bemoaned the influx of meat, dairy products, and animal fat into the region's diet, a phenomenon they called "Northernization." They decided to see what a return to a more traditional diet would do. Inviting 115 postmenopausal women to join a research study, 58 women were asked to cut back on meats, dairy products, and animal fat and favor plant-based foods, while the remaining women were asked to stick with their usual diets. Neither group increased their fiber intake. But, measuring estrogens excreted in urine samples, the researchers found that the women who cut back on animal products reduced estrogen levels by 40 percent.[6]

So the evidence is in: A higher-fiber, lower-fat diet brings a woman's hormones to a healthier level. While researchers are most excited about the reduction in breast cancer risk these changes can bring, they also help women struggling with other hormone-related issues.

Go Dairy-Free

Another threat to fertility may come from dairy products. This is not entirely surprising, because milk and cheese have plenty of unnecessary calories and lots of fat. Typical cheeses are *70 percent fat*, as a percentage of calories. So, as these products bring on extra weight, those growing fat cells contribute to hormone haywire. And, of course, dairy products are not plants, so they have zero fiber.

But dairy products have two more attributes that could aggravate hormone problems. First, in order to maximize their milk production, dairy cows are artificially inseminated every year. During their nine-month pregnancy, cows produce estrogens that get into

Lactose

milk. When milk is turned into cheese, these estrogen traces are more concentrated.

Second, the milk sugar, *lactose*, may be harmful to the ovaries. In 1994, Dr. Daniel Cramer at Boston's Brigham and Women's Hospital reported that the more milk women drink, the faster their decline in fertility as the years go by.

Normally, fertility declines bit by bit as women get older. In Thailand, where milk is not a traditional part of the diet, fertility declined by only about 25 percent as women passed from their late twenties to their late thirties.[7] In Finland, however, where dairy products are a major part of the diet, women had a much greater decline in fertility—around 80 percent—during the same age range. Statistics for New Zealand were similar, and Denmark, the United States, and the UK were not far behind.

Dr. Cramer reasoned that the problem may be lactose. During digestion, lactose breaks apart, releasing two smaller sugar molecules, glucose and galactose. It turns out that galactose can be toxic to the ovaries, damaging the germ cells that turn into eggs. There is plenty of lactose in milk and ice cream, ready to release galactose into your bloodstream.

D-galactose

You will find "lactose-free" cow's milk in stores, but that does not solve the problem. The fact is, it has just been treated so as to break the lactose into glucose and galactose in advance, so galactose is waiting for you right in the carton. Skip it. Plant milks are free of lactose naturally and are much better choices: soymilk, rice milk, oat milk, almond milk, hemp milk, and other varieties are now widely available. Most other plants are galactose-free, although traces do show up in dates, papayas, bell peppers, tomatoes, and watermelon.[8]

It Is Good to Be Lactose Intolerant

Many people are "lactose intolerant," meaning that milk sugar gives them digestive troubles. This is not a disease; it is totally normal for human beings (and all other mammals) to lose the ability to digest lactose in the years after weaning. After all, you are no longer nursing; you no longer need the *lactase* enzymes that break down lactose. Perhaps eliminating lactase enzymes is nature's way of protecting you from the dangers of lactose and galactose. If milk makes you sick, you won't drink it.

However, most Caucasians carry a genetic mutation that causes lactase enzymes to persist, a trait that is much less common in other races. That is no great benefit, because people who continue to ingest lactose into later childhood and adulthood are at risk for the toxicities that it can pose. Those include fertility problems and, as we will see in later chapters, a higher risk of ovarian cancer and prostate cancer.

Many people cite a need for calcium as a reason for consuming dairy products. Luckily, there are far more healthful sources. Green leafy vegetables and beans, in particular, are calcium-rich, as are many other foods from plants. By skipping dairy products, you will skip their fat, calories, cholesterol, estrogens, and galactose.

Polycystic Ovary Syndrome

A common cause of fertility problems is polycystic ovary syndrome, a condition marked by a group of symptoms often including ovarian cysts, irregular periods, and signs of excess male hormones, such as unwanted facial or body hair, acne, and male-pattern hair thinning.

There is a lot to say about this condition, so much so that I have devoted a whole chapter to it (Chapter 5, "Reversing Polycystic Ovary Syndrome").

Foods and Sperm Production

Making sperm cells is not an easy job. The body has to make millions of microscopic submarines, each one carrying the genetic plans for a human being, with the not-so-simple assignment of trying to find and fertilize an egg. Lots of things can go wrong in the process. Sometimes, there are not enough sperm cells, or sperm cells are not formed correctly or do not move very well.

Here again, dairy products come under suspicion. Researchers at the University of Rochester tracked the diets of 189 male college students—a group not known for exemplary eating habits—and they also checked their sperm samples. It turned out that men who consumed the most cheese and other fatty dairy products were more likely to have poorer sperm morphology and motility, compared with other men. That is, the shape and movement of their sperm were more likely to be abnormal. Their sperm counts tended to be lower, too. The researchers hypothesized that the estrogens in dairy products may have affected the men's reproductive biology.[9]

The researchers then visited a fertility clinic and checked the patients' diets. Men who ate cheese regularly had, on average, a 28 percent lower sperm concentration, compared with men eating less cheese or none at all. Their sperm motility and morphology were generally worse, too.

How much cheese were they eating? The low-cheese group averaged between zero and half a serving per day and the high-cheese group averaged between one and two and a half servings per day—not unlike what many Americans eat routinely.[10] It appears that even small amounts of cheese, if consumed on a daily basis, may disrupt male fertility.

A study in Spain found much the same thing: Men with poor sperm quality tended to eat more cheese and processed meat, and less fruits and vegetables, compared with men whose sperm were normal. The researchers speculated that chemicals in animal products could harm sperm production, while fruits and vegetables provide antioxidants that protect developing sperm cells.[11]

One might think that the traces of hormones or chemicals in dairy products would be too small to affect our health. But remember, dairy cows are pregnant nine months out of every twelve, and the amount of estrogen rises as the cow's pregnancy progresses. As I mentioned earlier, the process of turning milk into cheese concentrates the hormones.

In 2016, researchers systematically reviewed published studies on foods and male fertility, finding several studies indicting fatty dairy products and meat, especially processed meat (e.g., bacon, turkey bacon, sausage, ham, hot dogs, etc.), while finding benefits from fruits and vegetables. Among supplements, there seemed to be some benefit from omega-3s (eicosapentaenoic and docosahexaenoic acids, 1.84 grams per day), carnitine (L-acetylcarnitine and/or L-carnitine, totaling 3 grams per day), and coenzyme Q_{10} (300 mg once or twice per day).[12]

Using Exercise to Boost Fertility

Earlier, I mentioned that Harvard researchers had found that body weight affects fertility—it pays to trim away excess body fat. The Harvard researchers also looked at exercise.[13] Moderate exercise, such as walking at a regular pace, did not help fertility at all. But vigorous exercise was associated with *better* fertility, even among women with the same body weight. For every hour the women devoted to vigorous exercise each week, the risk of infertility dropped by 5 percent. So five hours of vigorous exercise each week cuts the risk by 25 percent. For running and jogging, the improvement was even greater.

Reduction in Risk of Infertility per Hour of Exercise per Week	
Aerobics:	5%
Biking:	5%
Lap swimming:	5%
Racquet sports:	12%
Jogging:	22%
Running:	34%

Rich-Edwards JW, Goldman MB, Willett WC, et al. Adolescent body mass index and infertility caused by ovulatory disorder. *Am J Obstet Gynecol.* 1994;171(1):171–177.

Does that surprise you? One might fear that vigorous exercise would disrupt a woman's monthly cycle. And indeed, researchers surveyed 394 women running in the New York City Marathon, finding that about one in four had reduced frequency of periods, or no periods at all.[14] But it turned out that the loss of periods was not actually associated with how vigorously the women exercised; it was related to excessive thinness. So if a woman exercises regularly but keeps her weight within a healthy range, fertility is unlikely to be affected and may well be improved.

Morning Wellness

Earlier, I shared Katherine's story of recovery from endometriosis and how she was able to avoid a hysterectomy and raise a family. She told me about one more experience that I would like to share with you. During her third pregnancy, something hit her that had not been a problem with her first two pregnancies. She had morning sickness. "My nausea was insane, practically making me immobile some days," Katherine said.

"Morning sickness" is really a misnomer, because, for many women, the nausea and vomiting hit any time of day. It is very common, but Katherine wondered why she had been free of it in her first two pregnancies and was suffering so much now. She took another look at her diet and realized that, although she had cured her endometriosis with a healthy, low-fat vegan diet, she had recently become more casual in her food choices. Theoretically, she was still avoiding animal products, "But at a restaurant, if the cook put cheese on my spaghetti marinara, I wouldn't take it off. Sometimes I had a greasy muffin or piece of cake that someone brought to the office. And I was having a lot of fried foods." By most people's standards, her diet was not too bad. Could these small-seeming dietary indiscretions be triggering her symptoms?

She decided to get back to a cleaner diet. No more "accidental cheese." No more animal products at all, and no more fried foods. Everything was vegan and modest in fat. And it worked. Within thirty-six hours, her nausea disappeared.

Katherine's experience fits a recommendation from the American College of Obstetrics and Gynecology for women who suffer from severe nausea and vomiting: The "BRATT" diet—bananas, rice, applesauce, toast, and tea—can be very helpful. And of course, it is low-fat and entirely plant-based.[15]

Her experience also fits with a theory proposed in 1976 by Ernest B. Hook, a researcher at the University of California at Berkeley. Dr. Hook pointed out that pregnant women often have aversions—to coffee, alcohol, or tobacco, for example—and are prone to nausea and vomiting. He suggested that, rather than being a "sickness," these are natural mechanisms for protecting both mother and baby.[16]

Normally, a woman's immune system protects her from pathogens. But during pregnancy, her immune defenses are naturally weakened to prevent her from rejecting the developing embryo. So nature adds another layer of defense: It makes a woman much more sensitive to the smell and taste of potentially tainted foods, so that if she has eaten something unhealthy, she will throw it up.[17] Indeed, morning sickness is especially acute in the first trimester, when the baby's organs are first developing and are most vulnerable to harm.

It turns out, from studies across many cultures, that the most common food aversion among pregnant women is meat. Some societies even have traditions or taboos against meat-eating by pregnant women.[18] This makes sense. Meat often harbors disease-causing germs: salmonella, *Escherichia coli*, toxoplasma, listeria, and many others. Researchers also observed that in cultures whose diets are based on corn and green vegetables, rather than meat, morning sickness is less likely to occur.[19]

None of this is an accident. It is all driven by hormones. As pregnancy begins, progesterone—a pregnancy-related hormone—weakens the immune defenses. It does this by reducing the number of specialized white blood cells, called *natural killer cells*, that attack invaders.[20] While it may sound strange to consider a developing baby an "invader," half the baby's DNA came from the father. To the mother's immune system, that means the baby is largely foreign. So progesterone weakens the immune system to ensure that the baby will not be rejected.

So it looks like morning sickness may be nature's way of steering you away from foods that could hurt you and your baby and toward foods that will support you both. Harvard researchers compared the diets of forty-four women who had had serious morning sickness with the diets of eighty-seven women who had had little or no nausea during pregnancy. It turned out that the two groups had noticeable differences in their diets. An especially noteworthy culprit was saturated fat—the "bad" fat that is found in cheese and other dairy products and in meat. Every 15 grams of saturated fat in a woman's diet (the amount in 2.5 ounces of cheese) increased the risk of morning sickness fivefold.[21] Putting it the other way around, cutting out dairy products and meat ought to more or less eliminate morning sickness, which is exactly what happened for Katherine.

Understanding how foods could trigger her symptoms was life-changing. "It set me free!" Katherine said. "It totally changed the rest of my pregnancy." She wanted to shout this from the mountaintops. "Most women just think nausea is inevitable. But there is an answer!"

Foods for Fertility

Planning for pregnancy and childbirth can bring up some complicated issues. But with the right foods on your side, the biological aspects become much simpler. Both Elsa and Katherine benefited enormously from a plant-based diet. That makes sense, because a healthy, low-fat, plant-based diet allows you to:

1. Trim excess body fat, a source of unwanted hormones.
2. Keep excess hormonal activity in check, through SHBG.
3. Have plenty of fiber to eliminate excess hormones through the digestive tract.
4. Avoid dairy products and the fat, calories, hormones, and galactose they harbor.

The same healthful foods that put hormones into better balance for pregnancy to occur are also especially friendly during pregnancy. The healthiest eating plans of all are those that skip animal products and focus on four healthy food groups: fruits, vegetables, whole grains, and legumes (beans, peas, and lentils). We will see how to put this into practice and how to ensure optimal nutrition in Chapter 12.

CHAPTER 2

Curing Cramps and Premenstrual Syndrome

I was sitting at my desk when the phone rang. The young woman on the other end of the line needed help. Her name was Robin, and she was suffering from severe, disabling menstrual cramps. Many women have cramps. But for some, the pain is off the scale. That was Robin's situation: For a day or two every month, her pain was so bad she could not function. This had been going on ever since her periods started, and nothing had ever gotten the pain under control.

On this particular day, she had been in agony, curled up on the floor of her office when her boss happened to walk past. Robin was scheduled for an out-of-town trip the next day, and this was obviously not going to work. Robin looked up. "I'm fine," she said unconvincingly. Her boss said, "No, you are not fine. Let's get you some help." So Robin called her mother, a physician, who had had exactly the same problem when she was younger.

Robin recalled how, when she was ten or eleven years old, her mother took her to the movies to see *Wuthering Heights*, one

21

of her mother's favorites. But partway through the film, her mother began shifting uncomfortably in her seat. Pain had gripped her, and they had to leave the theater. At thirty-five, her mother had a hysterectomy.

Robin's mother told her that the only thing that had worked for her was Demerol, a narcotic often used for postsurgical pain, and she encouraged Robin to call me. If I could give her a prescription to hold her for a day or two, perhaps she could function well enough to get on a plane.

As she spoke, I started to think through what menstrual cramps are and what could ease her misery. It occurred to me that, for years, researchers have looked at the connections between foods and hormones that put women at risk for breast cancer. We have long known that food choices can increase or reduce the amount of estrogens—female sex hormones—in a woman's bloodstream, as we saw in the last chapter. This is important because estrogens fuel the growth of cancer cells. They also cause the monthly changes in the uterus. If diet changes can calm the hormonal storms that fuel cancer growth, maybe they can do the same for cramps.

I let her know that I would be happy to give her painkillers for a couple of days. But I also suggested something else. "Would you like to try a diet change to see if we can prevent this from happening next month?"

"I'll try anything," she said.

So I described some simple food advice that aimed to reduce her hormonal changes over the next month: no animal products, and keep oils to a bare minimum. She agreed to give it a try.

Four weeks later, she called again. "This is amazing!" she said. Her period had sneaked up on her with no symptoms at all. No pain, nothing. Over the next several months, she felt great. She stuck with the dietary approach, and the pain that had arrived every month without fail finally gave her a reprieve.

Later on, she loosened her diet. She had gone home for the winter holidays and ended up eating more or less whatever was around. The next month, her menstrual pain hit so severely she could not go to work. It became clear that the diet change really worked and that if she deviated from it, she paid a price.

In fact, there were times when she wandered away from a healthy diet with some fattier foods, causing the pain to recur. Eventually, a gynecologist diagnosed endometriosis and suggested laparoscopic surgery to remove the overgrowing tissues that were causing the pain. The idea was appealing, because it meant that perhaps she could tackle her pain without needing to be so careful about food. So she decided to have the surgery. But it did not help. Her cramps were as bad as ever. The only things that had worked for her were narcotic pain medications and a diet change. She did not like the way she felt on pain medications. Her only answer was a healthful diet. And that worked like a miracle.

Foods and Menstrual Pain

Let me describe how foods might play a role in menstrual cramps:

At the beginning of a woman's monthly cycle, just as menstruation begins, there is very little estrogen in the bloodstream. Over the next two weeks, the amount of estrogen (especially estradiol) rises to a peak and then quickly falls. This is ovulation—the ovary is releasing an egg.

Then, over the next several days, the amount of estrogen rises again. The reason is that the uterus is the most optimistic organ in the body. Every month, it is convinced that a blessed event is about to occur, so estrogen thickens up the uterine lining (the endometrium) in anticipation of pregnancy. After about a week, however, the disappointed uterus discovers that it is not pregnant. The amount of estrogen falls rapidly, and the uterine lining

Estrogen Changes in the Monthly Cycle

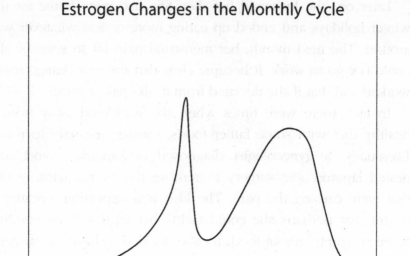

Menses   Ovulation   Menses

disintegrates in menstrual flow. As this lining breaks apart, it releases maladjusted chemicals, called *prostaglandins*, that cause crampy pain.

And here is where foods come in. As you know from the previous chapter, foods can boost estrogens (if you haven't read that chapter, please do. It's short, and it applies very much to this one). In a nutshell, a diet based on animal products (meat, cheese and other dairy products, and eggs), oily foods (fried foods, oily dressings), and fiber-depleted foods (white bread) boosts estrogen levels. If there is too much estrogen in the bloodstream, the uterine lining thickens up more than it should, creating more prostaglandins and more crampy pain at the end of the month. By reducing the amount of estrogen in the bloodstream, there is less thickening of the uterine lining, less prostaglandin production, and less pain. At least, that was my theory that led me to suggest a diet change to Robin, and it seemed to work.

I wrote about this at the time and began to hear from other women who tried this method and had similar success. So, my research team decided to put it to the test, and we published our findings in 2000. Working with researchers from Georgetown University's Department of Obstetrics and Gynecology, we invited women suffering with menstrual pain to join a research study. Half the women were asked to make some diet changes, setting aside animal products and avoiding added oils. To make this easier, we gave them recipes and tips for healthful choices at restaurants. The other half were given a "supplement" pill that was, in essence, a placebo—a dummy pill. Then, after two cycles (two months), the women switched; the women who had been following the special diet switched to the supplement, and the women taking the supplement switched to the diet.

Along the way, we asked the women how many days of pain they experienced. Before the study began, the women averaged 3.9 days of pain. The supplement did not do much of anything. But on the special diet, the pain was shortened to 2.7 days. We then asked them how bad the pain was, on a scale from 1 to 10. Before the diet change, the pain averaged 7 on the first day, 5 on the second day, and 3 on the third day. On the diet, they rated the first day of pain as a 6, on average, showing that many still had some pain. But the second day was a 3, and by the third day it was gone. The benefits varied from one person to another. Some had a complete disappearance of pain, others got partial relief, and a few found very little difference. But overall, the effects were very noticeable. These findings were published in the journal *Obstetrics & Gynecology*.[1]

We also saw a difference in their blood tests. We measured sex hormone–binding globulin (SHBG), the special protein molecule that circulates in the bloodstream, holding on to hormones like an aircraft carrier holds on to fighter planes, as I mentioned in Chapter 1, "Foods for Fertility." On the healthful diet, SHBG levels went up 19 percent. That change would be expected to calm hormonal effects.

The study was what researchers call a "crossover" trial—a study in which all the participants test two different treatments—diet and supplement. Some got the diet first; others got the supplement first. But when the time came for the diet group to go back to their old way of eating and begin the supplement, many refused. They felt so good, they did not want to return to the eating habits that had caused their pain. No amount of pleading on our part about the need to stick to the research protocol could induce them to stop what had seemed like a life-changing diet.

Anna

Anna joined a separate research study that focused on weight loss, not expecting that anything would happen to her menstrual symptoms. Three years earlier, she began having erratic and irregular cycles. "I was miserable most of the time from cramping, bloating, bleeding, sanitary concerns, odors, and fear of something worse."

Her bleeding was so irregular and heavy that she often wore a pad before leaving the house, just in case, and carried several types of pads and tampons, along with a change of clothes. She kept towels and cushions under her chair at home, at work, and in her car, and slept with towels layered underneath her sheets. She was humiliated at work when she accidentally stained her desk chair. She and her girlfriends at work began referring to menses as "red chair week." "I once had to wait until everyone had left before I could safely get up and move out of the room to clean up."

A medical evaluation turned up one or two fibroids, but nothing else. Her doctor advised her to try birth control pills, to see if they might help. But she decided against them. Joining our study, she began a low-fat vegan diet, aiming to lose weight. And she did, about nine pounds over sixteen weeks. But something else happened. "The first month of the vegan diet I had no cramps, bloating, or pain," she said. "My period was light and shorter in duration. It was really surprising. The second month was the same as the first. My coworkers and I celebrated the end of red chair days!"

Another participant in the same study reported a similar experience. After changing her eating habits, she lost weight and felt great, and, to her surprise, her periods were much shorter and more manageable.

Mary-Ann

Mary-Ann grew up near Cape Town, South Africa. Starting at age thirteen, she knew about hormonal issues firsthand. "I had heavy periods lasting five to seven days, with huge clots, strong odor, and bad pelvic and back pain, plus premenstrual tension," she said. "I was pale, tired, and in pain, and I often soiled the sheets no matter how many sanitary towels I used."

Her family was reasonably health-conscious for that time. Her mother served whole-grain bread and would not allow her children to have coffee or tea. Even so, her grandfather had been a sheep farmer, and the family tradition included plenty of lamb and chicken, with barbecues every weekend. And she developed a real love for cheese. "I was a cheese addict," she said.

When Mary-Ann was twenty-nine, her baby girl—the youngest of three—began to have frequent ear infections. Because she was breastfeeding, Mary-Ann began to wonder if somehow her own diet was contributing to her baby's problem. So she decided to stop eating dairy products altogether, and that did indeed help her baby. It also helped her. Her periods became noticeably shorter and lighter, with less odor.

Later on, she eliminated animal products altogether. "And within three months, my period was just two to three light days, with no pain at all, no heavy blood clots, no odor, and no premenstrual symptoms!" This was amazing.

Her husband joined her in this new way of eating, and they raised their daughters on healthful plant-based diets, too. As their daughters reached adolescence, the girls had very light periods with no pain. Their school friends found this surprising, telling them it was not normal to have periods lasting only two days, with no pain at all.

When their daughters married meat-eating men and started eating animal products, they developed heavy, painful periods. "My daughter called me to ask what was wrong and why she was in such pain. I encouraged her to change her diet back. She did, and her pain went away."

Today, Mary-Ann loves a cold mango on a hot day. She loves asparagus, broccoli, and the full range of vegetables. She launched a vegan restaurant, called Mary-Ann's Natural Emporium & Eatery, and serves a wonderful zucchini lasagna made with grated butternut, sliced artichokes, leeks, cashew milk, and tomato puree. And she set up a basic nutrition course, helping people improve their diets. In the course of her work, she has seen menstrual disorders disappear for many women.

Many women suffer with cramps and have never been told about what a diet change could do, because their doctors do not know about it. That should change. Medications and surgical options that are used to treat these conditions have risks. But the "side effects" of healthful foods are all things you want—weight loss, improved heart health, reduced cancer risk, and better health overall, in addition to hopefully knocking out your pain.

Endometriosis

In Chapter 1, we touched on endometriosis and how Katherine conquered this problem, avoided surgery, and was able to raise a family. Endometriosis is when cells that are supposed to be lining the uterus migrate to other parts of the abdomen—the ovaries, fallopian tubes, or elsewhere—where they implant, causing pain and, often, infertility.

The usual treatments are pain medications, birth control pills, other hormonal treatments, and, in some cases, surgery. But, as Katherine found, a change in your diet can be just what the doctor ordered. By focusing on plant-based foods, skipping animal products entirely, and minimizing the use of added oils, she cured her endometriosis.

Fibroids and Adenomyosis

Inside the wall of the uterus is a layer of muscle that allows the uterus to contract. In many women, these muscle cells overgrow, forming a knot, called a *leiomyoma*, more commonly known as a *fibroid*. Most women have small fibroids by age fifty, and often they present no symptoms at all. Sometimes, however, they become large and painful and can cause heavy menstrual bleeding. They are the most common reason for a hysterectomy in the United States.

Adenomyosis is a condition in which cells that normally line the uterus have ended up embedded in the uterus's muscle layer. Like fibroids, they are common and sometimes cause no symptoms at all. But they can cause menstrual pain, heavy flow, and pain during intercourse.

The good news is that fibroids and adenomyosis are not cancerous, and both conditions improve on their own after menopause. However, you might not want to wait that long. Here's what you need to know:

Fibroids and adenomyosis are fueled by estrogens. That is why they shrink after menopause, as estrogen levels decline. But that also means that the steps that tackle estrogen can help you *now*, at least in theory.

Let's start with weight loss. Body fat makes estrogens. So trimming away excess body fat helps. The situation is similar to infertility, as we saw in the last chapter, where it pays to be a bit on the thin side. Just as modest increases in body fat make infertility more likely, the same is true for fibroids. This does not mean one should be overly thin. But a BMI in the range of 19 to 20 kg/m² is where fibroids seem to be least frequent.[2,3,4] For a woman who is five feet five inches tall, that would be 114 to 120 pounds. Even if you don't quite get there, losing excess fat is always a good idea.

The healthiest way to lose weight is with a low-fat, plant-based (vegan) diet. As we will see in Chapter 12, "A Healthy Diet," these foods are naturally modest in calories and give you a metabolic boost, so that you can lose weight without trying to starve off the pounds or forcing yourself to say no to carbohydrates, the way some old-fashioned diets want you to do.

Vegetables, fruits, whole grains, and beans also provide healthy fiber. As you will remember from the previous chapter, fiber helps your body eliminate unwanted estrogens. In contrast, animal products have no fiber at all, and dairy products *add* estrogens to your body, something you don't want.

There may be special benefit to eating plenty of vegetables and fruits in particular. Three studies—one in Italy, one in the United States, and one in China—found that women who eat plenty of fruits (particularly citrus) and green vegetables have a surprisingly low risk of fibroids.[5] Researchers have suggested that the benefit may come from vitamins, minerals, and other compounds naturally found in these healthful foods, but exactly where the credit goes is not yet certain.

Premenstrual Syndrome

Premenstrual syndrome (PMS) is a combination of physical and emotional symptoms that arrive in the days before your period. When the mood symptoms are especially prominent, doctors call it *premenstrual dysphoric disorder.* "Dysphoria" is the opposite of euphoria—it means you feel rotten, with mood swings, irritability, anger, depression, and anxiety. You can lose interest in your usual activities and have poor concentration, a lack of energy, and changes in appetite and sleep, among other symptoms. Then, around the time that menstruation begins, all these symptoms evaporate. What is apparently happening is that the hormone shifts, along with the prostaglandins released from the uterine lining into the bloodstream, cause all manner of physical and psychological symptoms.

In the research study I mentioned earlier, we asked our participants not only about pain, but also about how many days each month they experienced premenstrual symptoms—both physical and psychological. It turned out that water retention was cut from 2.9 days at the beginning of the study to 1.3 days on the diet. And what we called "behavioral changes," referring to moods and emotional reactions, were cut from about 1.7 days to 1.1 days on the diet. For many, the symptoms were not only briefer, but also milder. In the bargain, they lost weight, their cholesterol levels improved, and they had more energy. Instead of feeling stuck with recurrent pain and other symptoms, many gained a new level of control over their health.

Why would a plant-based diet help PMS? We do not know for sure, but a couple of possibilities present themselves: For starters, it reduces the amount of estrogen in the bloodstream. Some have suggested that that smooths out the ups and downs of your estrogen roller coaster, stabilizing your mood. Less estrogen also means less thickening of the uterine lining (the endometrium). That is important, because the endometrium is like a factory producing prostaglandins—those mean-spirited compounds that not only cause crampy pain but also circulate in your blood, making you feel crummy. If the endometrium does not thicken up so much, it cannot produce the tidal wave of prostaglandins that have been bothering you. And suddenly, you feel like yourself again.

I should mention that there are actually several different prostaglandins. Some are the malicious ones that I mentioned above. Others are trying to be kind to you. These helpful anti-inflammatory prostaglandins are trying to cool your symptoms. To boost the helpful ones, some people recommend foods rich in *alpha-linolenic acid* (ALA). ALA is an omega-3 oil, sometimes called a "good fat." You will find it in many plant foods, especially walnuts, ground flaxseeds, and soy products, among others. However, my best guess is that you will find greater relief by throwing out animal products and keeping oils low overall, because it will keep your prostaglandin "factory"— your endometrium—from thickening too much.

Give It a Try

If you would like to get into better balance and tackle pain or PMS, you can easily put foods to the test for yourself. Follow an entirely plant-based diet without added oils for at least one or two full monthly cycles—starting from the beginning of your period. It does not work if you do it intermittently or adopt a healthier diet for just a week or two before your period. You'll want to give it a good test. You are likely to feel much better, and the same diet changes that tackle cramps and PMS also improve your health in other ways, as we have seen.

While you are at it, you might wish to minimize or avoid caffeine and alcohol. Both can disrupt sleep, making daytime moodiness worse. Exercise has the opposite effect—it helps you sleep and can be a mood-booster.

The Power of Plant Protein

For an extra mood-stabilizing effect, you might try having a serving of plant protein as you start your breakfast. Good choices include grilled tempeh, scrambled tofu, veggie bacon, veggie sausage, soymilk (most other plant-based milks are low in protein), beans, or chickpeas. A little bit is all it takes. The idea is to have one of these plant-protein foods first, with starchier foods (e.g., toast, oatmeal, or fruit) later in the meal. To be clear, it is fine to eat starchy foods, but have the high-protein food first.

Some women have reported that it helps their moods for the rest of the day. This anecdotal observation has held up for people who have repeatedly tested it for themselves and found a noticeable "anti-grouchy" effect.

Some women report that sugar and chocolate may have the opposite effect, increasing moodiness. See if this is true for you.

In the evening, you can do the opposite: Lower-protein, higher-carbohydrate foods (e.g., bread or spaghetti) may help you sleep more soundly.

Two caveats: First, I do not mean to suggest that you are in any way protein-deficient or that you have some nutritional need for extra protein when on a plant-based diet. The fact is, any normal variety of grains, beans, and vegetables eaten throughout the day gives you all the protein you need for good health. The idea of boosting plant protein at breakfast is simply to take advantage of a mood-stabilizing effect. If it works for you, great. If not, do not feel that you need to boost protein for any nutritional reason. Second, I would advise against using eggs, bacon, sausage, or other animal-derived products as a protein dose. For the reasons that we've discussed and for those that will follow, they will do more harm than good.

Easy Steps for Tackling Pain and PMS

Our goal is to get hormones into better balance. We'll do that by basing the diet on fiber-rich plant foods that will help the body eliminate excess estrogens and by avoiding dairy products, which contain estrogens from the cow. In studying menstrual cramps and PMS, some researchers have also found benefit from vitamin D, vitamin E, calcium supplements, omega-3 supplements, or natural progesterone cream. But food is the key. It is what gets your body's chemistry into better balance.

So, to prevent or reduce menstrual pain and PMS:

1. Avoid animal products completely. This means meats, dairy products, and eggs.
2. Have plenty of high-fiber foods, like beans, vegetables, fruits, and whole grains. You will want to choose brown rice instead of white rice and whole-grain bread instead of white bread.
3. Keep oils low. It pays to avoid oils in cooking. Instead, use non-oil cooking methods, like steaming, boiling, baking, or sautéing in vegetable broth. Choose oil-free dressings on salads, and skip fatty foods, like French fries, potato chips, greasy pastries, and the like, and minimize nuts, nut butters, and avocados.
4. Take advantage of plant-protein-rich foods, like tofu or tempeh, especially as you start your breakfast.
5. Avoid sugar and chocolate. The effects are very individual, but if these foods make you irritable, steer clear.
6. Minimize caffeine and alcohol.
7. Take vitamin B_{12} daily. We'll talk more about vitamin B_{12} in Chapter 12. For now, suffice it to say that it is critical for brain function. Adults need only 2.4 micrograms per day, and inexpensive supplements are available at all drugstores and health food stores.
8. Get regular exercise. The Physical Activity Guidelines for Americans, developed by the U.S. government, recommend two and one-half to five hours of moderate-intensity physical activity every week. If you prefer, substitute 75 to 150 minutes of vigorous activity each week.
9. Get plenty of sleep.
10. Get sunlight. Sunlight is a mood-booster. Sunlight on your skin gives you the vitamin D you need. You will find it also helps your mood.

Do this 100 percent. Do not have a bit of chicken or yogurt here or there, thinking that it doesn't matter. Even small amounts of these foods can set you back, and that is true even when you have them weeks before your period.

Sound challenging? It might feel that way at first. But you will be surprised at how easy it soon becomes and at how many foods fit the bill. Some foods have pretended to be your friends all these years, and you've finally come to see them for what they are. Meanwhile, you will discover healthier foods that have been adoring you from afar.

Lindsay

As I was preparing this book, I asked Lindsay Nixon to devise the recipes. Lindsay is the author of *The Happy Herbivore Cookbook* and many other excellent books and is a master at making healthful eating approachable and engaging. As you will see in the recipe section, Lindsay came through beautifully. The recipes are truly wonderful.

But Lindsay shared with me that this book had personal meaning for her, too. "I suffered from debilitating menstrual cramps," she wrote. "I was always missing school and work. I suspect I had endometriosis. My OB/GYN wanted to put me on artificial hormones and painkillers, which scared me at age twenty-one to twenty-two. I refused and just suffered with it. But at age twenty-four, I went vegan. Within a few months, most of my symptoms lessened or disappeared. I haven't had a single pain in eleven years. Suffice it to say this is something very personal to me, and I am so glad you are writing this book."

Pain-Free and Loving It

If cramps and PMS have been lying in wait for you every month, see what a diet change can do. For many women, it is life-changing. Even better, the foods that tackle these monthly problems also promote lasting weight control and help you prevent serious health problems down the road, as we will see in the next chapter.

CHAPTER 3

Tackling Cancer for Women

We have new power to reduce our risk of developing cancer. And for people diagnosed with cancer, we have better methods than ever to improve survival. In this chapter and the next, we will zero in on cancers where the major issue is hormones. For women, that means breast, uterine, and ovarian cancer. For men, it means prostate and testicular cancer. For each of these, your risk can be changed by food choices, opening up exciting possibilities for reducing cancer risk and for enhancing survival in people who have already been diagnosed.

Lee

Lee grew up in northern Virginia, just outside Washington, DC. In college, she was healthy and active and was busily studying biology. But she began to notice pain in her breasts. It was mild at first, but over time it worsened. And it waxed and waned with her monthly cycle.

She went to see a doctor, who gave her a diagnosis—fibrocystic breasts—but was not able to recommend any treatment that would fix the problem. Apparently, it was something she would have to live with, and hope for the best.

By age thirty, continuing pain led her doctor to recommend a mammogram, which turned out to be not entirely normal. It showed microcalcifications in an arrangement that suggested a one-in-twenty chance of cancer. A biopsy was scheduled that thankfully did not show cancer, but did reveal changes that put her at higher-than-average risk in the future. So she made an appointment with a breast surgeon to see what she could do to stay healthy. The surgeon examined her and found a new area of thickened tissue. They measured it and decided to check it every three months for a year to make sure it was stable.

None of this was very encouraging. So Lee dug into the Internet. Breast cancer, she found, was linked to Western diets—that is, the meaty, cheesy American diet she was used to eating. She also realized that these foods might be contributing to her chronic pain.

She resolved to bring on healthier foods. She minimized animal products, alcohol, and processed food, and built her diet around plant-based foods. Her surgeon found no changes in the thickened tissue at her three-month, six-month, or nine-month checkups.

Then Lee slipped off the wagon. She brought some meaty, sugary, higher-fat foods back into her diet. After four months of this, her surgeon did what should have been her final checkup and sonogram. However, in that four-month period, the lump had doubled in size. The surgeon performed a lumpectomy and found what was called *flat epithelial atypia*, a condition that can be a prelude to cancer. The surgeon excised the growth, and hopefully, that was it.

For Lee, this was a serious shot across the bow. She knew that she was at risk for cancer and resolved to make big changes in her eating habits. Out with the animal products, in with the vegetables, fruits, whole grains, and beans, and no more cheese sauce or butter on veggies.

This time, she made it stick. Her diet change turned out to be a good move in many ways. She lost unwanted pounds and ended up at her high school weight. Her recurrent breast pain loosened its grip, and her annual checks have been all clear. So far, so good.

Her husband joined her in the menu rethink and, in the process, trimmed away thirty-five pounds and cured his borderline high blood pressure. Her mother joined her, too, and over time ended up losing a hundred pounds.

Lee has continued with her healthy vegan diet. And she went a step further. She followed her newfound interest in food and health and decided to become a registered dietitian. Today, she helps other women and men improve their diets, just as she did.

Tackling the Hormones That Cause Breast Cancer

Normally, the cells of the breast—like other organs—stay put, more or less. Trouble starts when something damages the chromosomes—the DNA blueprints deep inside the nucleus of the cell. When they are damaged, the cell can lose its ability to behave normally. It starts multiplying over and over, creating a growing tumor that can invade nearby tissues. Pieces of the tumor can break away and travel to other organs, where they invade.

Breast cancer is the most common cancer women face, after skin cancer. For the most part, it is not genetic. In perhaps 5 to 10 percent of cases, genes play a major role, and they play a smaller role in others. A big factor in whether breast cancer strikes or not is hormones—that is, estrogens.

It sounds odd to suggest that estrogens could cause cancer, doesn't it? After all, they are normal female sex hormones. Your body makes them, and they have important biological roles. Without estrogens, women would not have a monthly cycle. They would not have babies. But when things do not go right, estrogens are trouble.

Arriving at a breast cell, estrogens can sometimes do all kinds of mischief. They can sneak through the cell membrane. They can even penetrate the nucleus and damage your DNA, turning a healthy cell into a cancer cell.

And they don't stop there. Once a cancer cell arises, estrogens can stimulate it to multiply. Like fertilizer on weeds, they create a growing tumor that can invade and spread.[1]

Postmenopausal women with higher levels of *estradiol* in their bloodstreams have more than double the cancer risk, compared with women with lower levels.[2] Especially problematic is *free estradiol*—that is, estradiol that is not bound to carrier proteins

Breast Cancer Risk at Increasing Free Estradiol Concentrations

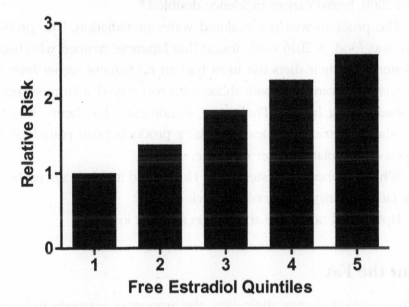

Endogenous Hormones and Breast Cancer Collaborative Group. Endogenous sex hormones and breast cancer in postmenopausal women: reanalysis of nine prospective studies. *J Natl Cancer Inst.* 2002;94:606–16.

in the bloodstream. As its name suggests, it is free to enter breast cells and do its work, for good or ill.

So what does all this have to do with food? Let's take a lesson from Japan. Over many centuries, Japan's food traditions were based mainly on rice and vegetables. Dairy products were not used at all, and when meat and fish were used, they were mostly just flavorings—the way Americans might use onions or pickles. Japanese people had the longest life expectancy on the planet.

For cancer researchers, Japan was a role model. Breast cancer was rare. And if Japanese women did get breast cancer, it was often less aggressive than cancer tended to be in American women, and they were more likely to survive.[3]

That all changed in the later decades of the twentieth century. Western eating habits began to invade Japan. Meaty business lunches became commonplace, and fast-food chains started to crop up, featuring burgers, chicken, and cheese. Between 1975 and 2000, breast cancer incidence doubled.[4]

The problem was not polluted water or radiation. The problem was food. A 2016 study found that Japanese women who had Westernized their diets the most had an *83 percent higher breast cancer risk*, compared with those who had stayed with more traditional eating habits.[5] The same phenomenon has been shown in other countries: As meat and dairy products push plant-based foods off the plate, cancer rates rise.

What is it about a Western diet? How could it cause cancer? Is it the fat? The dairy products? The lack of fiber?

How about "all of the above"? Let's have a look.

Cut the Fat

When women change their diets, the amount of estrogen in their blood changes, too. While high-fiber foods tend to reduce estrogen levels, fatty foods do the opposite. They increase estrogens.[6,7,8]

The same theme became clear in a Japanese government study that checked estrogen levels in 324 women. The more fat the women ate, the higher their estrone levels. This was true even among thinner women and for more or less all fats—the saturated fat found in milk and meat as well as the fats found in vegetable oils.[9] Higher hormone levels mean higher risk of developing breast cancer.

Fatty foods cause problems even after cancer has been diagnosed, evidence suggests. At the State University of New York in Buffalo, researchers studied women who had been diagnosed with advanced cancer.[10] They found that those who ate more fatty foods had a much higher risk of succumbing to the disease. Here are the numbers: An extra 1,000 grams of fat per month (about 30 grams of fat in a day) increased the risk of dying by about 40 percent. What does that look like in actual food terms? A typical 6-ounce serving of chicken has nearly 30 grams of fat, as does a 2-ounce serving of cheddar cheese and a 6-ounce serving of salmon.

If you were to add up the fat in a typical American diet and compare it to a plant-based diet without added fat, the two diets would differ by about 1,000 to 1,500 grams per month, which corresponds to a 40 to 60 percent difference in the risk of dying at any point in time. In other words, there is a lot of fat in a typical American diet, and getting away from it is a good idea.

The value of cutting fat intake was put to the test in the Women's Intervention Nutrition Study (WINS). It included 2,437 women who had previously been treated for breast cancer.[11] Some were asked to begin a low-fat diet. The others continued their usual diets. After five years, the risk of cancer recurrence was reduced by 24 percent in the low-fat group.

To avoid excess fats, the first step is to skip animal products. This helps you avoid the worst actors, the saturated fats in dairy products and meat.[12,13] In turn, your hormone levels adjust. In a large study called the European Prospective Investigation into Cancer and Nutrition, postmenopausal women following a vegan diet had estradiol levels that were 6 percent lower and sex hormone–binding globulin (SHBG) levels that were 19 percent higher, compared with meat-eaters. As you'll recall, SHBG is a helpful protein in your bloodstream that keeps hormones inactive until they are needed. These differences are modest, but in the right direction.[14]

A later study in Scandinavia found much the same thing: Vegetarians had less estrogen and more SHBG.[15] My research team confirmed that women who adopt a vegan diet quickly boost their SHBG levels. In five weeks, SHBG levels increased 19 percent.[16]

In addition to avoiding animal products, it also pays to favor cooking methods that do not use added oils. As you will see in Chapter 12, "A Healthy Diet," it is easy to shift to lighter, "cleaner" ways of food preparation. You'll not only be addressing cancer risk, but you'll also find a huge payoff in weight loss, lower cholesterol, and better overall health.

Dump Dairy

In the 1990s, a Japanese television crew came to the United States and interviewed me for a program on nutrition. The interviewer was about thirty, and when the subject of dairy products came up, she visibly wrinkled her face in disgust. Afterward, she explained that there had been a big push to increase milk consumption in Japan—to be more like Americans—and she found it disgusting. She had never gotten used to it, and she was not convinced that it was healthy.

She was right to be skeptical. As we saw in Chapter 2, "Curing Cramps and Premenstrual Syndrome," dairy products harbor estrogens. Dairy cows are impregnated annually, and as each pregnancy progresses, they make more and more estrogen. Although only traces end up in milk products, researchers are increasingly concerned that these traces may be enough to affect human health.

Australian researchers measured hormone levels in 766 post-menopausal women. Those who consumed the most dairy products had 15 percent more estradiol in their bloodstreams, compared with women consuming little or no dairy products.[17]

Dairy hormones may affect your odds of surviving breast cancer. A California study of women diagnosed with breast cancer found that those consuming one or more servings of whole-fat dairy products (e.g., butter and cheese) per day had a 49 percent increased risk of dying of their cancer over the twelve years of study, compared with those who generally avoided these foods.[18]

Although most of us grew up with the notion that dairy products were healthful, science has come to view them very differently as their effects on health have become clearer. You are far better off getting your calcium from broccoli, kale, collards, Brussels sprouts, or other green leafy vegetables. In the bargain, these healthful foods are rich in vitamins, with none of the fat, cholesterol, lactose, or hormones found in dairy products.

Boost the Fiber

Part of the value of fruits, vegetables, whole grains, and legumes is that they aren't cheese. Let's face it, you can't impregnate a strawberry, and it will never crank out estrogens. But there is more to it. Plants have fiber. And fiber escorts unwanted hormones out of your body. As we saw in Chapter 1, "Foods for Fertility," the liver filters excess estrogens out of the blood and sends them into the intestinal tract, where fiber carries them away. The more your menu is built from plant foods, the better this system works.

In 2011, researchers took out their calculators and figured out just how helpful fiber is. They estimated that every 10 grams of fiber in the diet cuts breast cancer risk by 7 percent.[19] So, a woman who boosts her fiber intake from 10 to 20 grams a day cuts her risk by 7 percent. Adding 10 more (going up to 30 grams) would cut it by 14 percent, 40 grams would cut it by 21 percent, and so on. Ten grams is the amount of fiber in a serving of beans, plus a fruit.

In Harvard's Nurses' Health Study, the findings were even better. Women who got 30 grams of fiber each day were 32 percent less likely to develop breast cancer, compared with women who got relatively little fiber.[20] Thirty grams is more than average Americans or Europeans are getting now, but it is easy to achieve. You can get this amount by having, say, bran cereal for breakfast, bean chili with brown rice and spinach for lunch, a piece of fruit for a snack, and lentil soup and spaghetti with tomato sauce for dinner. Once the cheese and meat are off your plate, there will be plenty of room for healthier choices. The key to remember is that plant products have fiber, and animal products do not.

Fiber Finder

Here is a simple guide to estimating the amount of fiber in typical foods. Different varieties will bring you slightly different amounts, but these are good overall averages to get you started. Aim for 40 grams per day.

Legumes

Beans, peas, lentils (½ cup): 7 grams
Soymilk (1 cup): 3 grams
Tofu (½ cup): 3 grams

Vegetables

Broccoli, carrots, or other common vegetables (1 cup, cooked): 5 grams
Lettuce (1 cup): 2 grams
Potato with skin: 4 grams
Potato without skin: 2 grams

Fruit

Apples, oranges, and other common fruits: 4 grams

Fruit juice (1 cup): 1 gram

Grains

White bread (1 slice) or bagel: 1 gram

Whole-grain breads (1 slice): 2 grams (brands vary; check labels)

White pasta (1 cup, cooked): 2 grams

White rice (1 cup, cooked): 1 gram

Brown rice (1 cup, cooked): 3 grams

Oatmeal (1 cup, cooked): 4 grams

Bran cereal (1 cup): 8 grams

Typical processed cereals (1 cup): 3 grams

Animal Products

Meat, poultry, or fish: 0 grams

Dairy products: 0 grams

Eggs: 0 grams

Beware of Carcinogens in Foods

Apart from their hormonal effects, meaty meals deliver carcinogens you never bargained for. Heterocyclic amines are cancer-causing chemicals that form as meat is cooked. In the United States, the biggest source is chicken. If that sounds surprising, Americans eat *one million chickens every hour.* So the carnivores among us swallow a lot of heterocyclic amines. They are in other meats, too.

Apart from chicken, among the worst of the lot are *processed meats.* Turkey bacon, pork bacon, sausage, hot dogs, ham, pepperoni—these foods cause breast cancer.[21] The more you

(continued)

include them in your routine, the higher your risk. In case you were wondering, plain old red meat is also implicated. All meats have a tendency to accumulate carcinogens as they cook, and they also have a tendency to encourage the formation of carcinogens in your digestive tract. None of this does your body any good.

There are other sources of carcinogens, too. We will have more to say about them in Chapter 13, "Avoiding Environmental Chemicals."

Fat Cells Are Hormone Factories

If we have not beaten up on animal products enough yet, they raise another important concern. Diets based on animal products tend to be fattening, compared with a traditional rice-based diet in Japan or a plant-based diet anywhere else. No, beef and chicken breast do not keep you lean. Even skinless chicken breast is about one-quarter fat, as a percentage of calories. Roast beef is higher, and cheese is higher still.

Fatty foods are calorie-dense and never have any fiber to help you stay slim. That is why meat-eating adults tend toward being overweight, while most people who avoid animal products are in the healthy weight range.

As meaty diets expand your fat layer, each fat cell produces estrogens, as we saw in Chapter 1. Have a look at the figure below. Researchers with the Women's Health Initiative drew blood samples from 267 postmenopausal women. They found that the more body fat women had, the more estrogens were in their bloodstream. This was true for both estradiol and estrone.[22]

One more thing: Heavier women tend to have less SHBG in their bloodstreams. So body fat is a double whammy. This means extra estrogen molecules circulating in your blood, and without SHBG to rein them in, those hormones are more active and more likely to contribute to cancer risk.

Body fat does the same thing to men. Next time you are at the beach, look around. Overweight men often have breast

More Body Weight Means More Estrogen

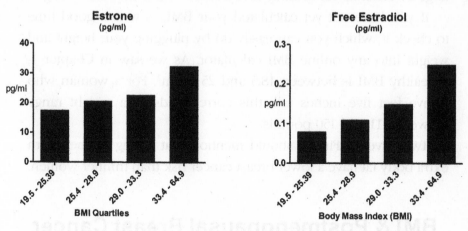

McTiernan A, Wu L, Chen C, et al. Relation of BMI and physical activity to sex hormones in postmenopausal women. *Obesity.* 2006;14:1662–77.

enlargement. This is not just fat. Their body fat is producing female sex hormones—estrogens—which in turn cause the formation of breast tissue.

All of this helps explain what happened in Japan. As meat and cheese displaced traditional rice and vegetables, they delivered carcinogens to Japanese meals and also started expanding Japanese waistlines, so fat cells could produce more and more estrogens, ready to increase cancer risk. And the rest is history.

So, does slimming down help? The answer is yes. It helps a lot. In addition to all its benefits for health in general, weight loss reduces your odds of developing postmenopausal breast cancer.[23] And that is true no matter how you lose it—through a healthy eating plan, surgery, or other means. Of course, there are major advantages to doing it through foods, but the point is that shedding excess body fat is a good idea.

Look at the graph below. Harvard researchers found that heavier women (with a BMI above 30 kg/m²) were 47 percent more likely to develop breast cancer, compared with those with BMIs under

23. But as your weight descends, so does your cancer risk.[24] A large Swedish research study found more or less the same thing.[25]

If you have not yet calculated your BMI, now is a good time to check it, which you can easily do by plugging your height and weight into any online BMI calculator. As we saw in Chapter 1, a healthy BMI is between 18.5 and 25 kg/m². For a woman who is five feet five inches tall, this corresponds to a weight range between 111 and 150 pounds.

Two caveats: First, I should mention that young women with extra body fat have a lower breast cancer risk than thinner women.

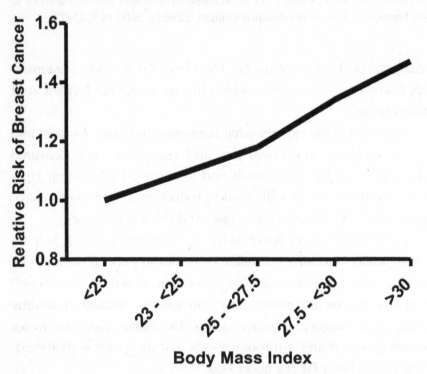

BMI & Postmenopausal Breast Cancer

Zhang X, Eliassen AH, Tamimi RM, et al. Adult body size and physical activity in relation to risk of breast cancer according to tumor androgen receptor status. *Cancer Epidemiol Biomarkers Prev.* 2015;24(6):962–968.

That lasts until menopause, after which the risk for overweight women is much higher than thinner women. Carrying extra weight into midlife is also linked to diabetes, hypertension, heart disease, and other health problems.

Second, thinner is better only up to a point. Women who are seriously underweight are at higher risk, too. "Underweight" means a BMI below 18.5 kg/m² (that would be a weight under 115 pounds for a woman who is five foot six).

Trimming away excess body weight is not just good for reducing cancer risk. It also improves survival in women who have been diagnosed with cancer. When cancer strikes, women at a healthy weight have better chances of survival, compared with overweight women.[26]

Even modest weight differences affect survival. A study in Shanghai followed a large group of women with breast cancer.[27] Those with BMIs above 25 had a five-year survival of 80.1 percent. For those with BMIs just slightly lower—between 23 and 25 kg/m²—survival odds edged upward to 83.8 percent. For those with a BMI below 23 kg/m², the five-year survival rate was 86.5 percent. You get the idea. As excess body weight is trimmed away, the odds of survival gradually increase, and this is true even within what many people would think of as a "healthy" weight range.

For women who are at an unhealthy weight when they receive a cancer diagnosis, it is not too late to benefit from healthful weight loss. In Chapter 12, we'll look at the best ways to lose weight and to keep it off long-term.

Foods That Help You Detox

Despite your best efforts, some mean-spirited chemicals will make their way into your body, ready to damage your DNA and cause cancer. Certain foods can help you eliminate them. Although this book focuses on hormones, let me share a few tips on this aspect of cancer prevention.

When you breathe in, you take in oxygen, which is essential for every cell in your body. Even so, oxygen molecules are unstable and can easily be damaged. They lose electrons, or electrons end up in unstable orbits. These altered oxygen molecules are called *free radicals*. They can attack your skin, your blood vessels, and even your DNA, leading to cancer.

Luckily, *antioxidants* can neutralize free radicals. They are easy to find. Here a few key ones:

Vitamin C patrols the watery parts of your body—your bloodstream and the interior of your cells—knocking out free radicals that it encounters. You will find it in citrus fruits, of course, but also in many vegetables.

Beta-carotene shields you from free radicals that reach the surface of your cells. Its orange color shows you where to find it. Carrots, sweet potatoes, pumpkin, and cantaloupe are loaded with it. You will also find it in kale and other green vegetables. I would encourage you to get your beta-carotene from foods, not from supplement pills. In research studies, beta-carotene pills lack the cancer-prevention power that beta-carotene-rich *foods* have.

Lycopene is the red color in tomatoes, watermelon, and pink grapefruit.

If you are undergoing chemotherapy or radiation, clinicians sometimes limit antioxidants during active treatment on the theory that they could inadvertently protect the tumor cells you are trying to knock out.[28] I strongly encourage you to speak with your health care provider about how antioxidant-rich your diet should be, and whether limitations apply only to antioxidant *supplements*, as opposed to vegetables and fruits.

In addition to these antioxidants, let me say a word of praise for *cruciferous* vegetables—broccoli, cauliflower, Brussels sprouts, cabbage, kale, and their botanical cousins. Their name comes from their cross-shaped flowers, and they have a special power. They help you eliminate unwelcome chemicals that could otherwise lead to cancer. Here's how:

Your liver is constantly filtering your blood. When a toxic molecule shows up, your liver's *phase 1 enzymes* grab hold of the toxic molecule. These enzymes attach an oxygen molecule to it, like a policeman attaching one end of a pair of handcuffs to a criminal's wrist.

Then, a phase 2 enzyme attaches the other end of the oxygen molecule to a large carrier molecule, such as glutathione, which acts like a big, burly policeman carrying the toxin away. Cruciferous vegetables stimulate the liver to produce extra phase 2 enzymes, giving you a huge squad of policemen to carry away toxins.

Let's say you were to have broccoli on Monday and Tuesday. By Wednesday, your liver would have an extra set of detoxifying phase 2 enzymes and would be more vigilant against chemical intruders. There are also some noncruciferous vegetables with a similarly helpful effect, as you will see in the table below.

Foods That Boost Detoxifying Enzymes

The following foods increase liver phase 2 enzymes that remove toxins from the bloodstream.

Asparagus	Ginger
Broccoli	Green beans
Brussels sprouts	Green onions
Cabbage	Kale
Carrots	Leeks
Cauliflower	Lettuce
Celery	Spinach

Testing the Power of Fruits and Vegetables for Cancer Survival

The power of vegetables and fruits was put to the test in the Women's Healthy Eating and Living (WHEL) Study.[29] The goal was not to prevent cancer, but rather to help women who already had been treated for breast cancer. The study included 3,109 women. Half were asked to have five fruit and vegetable servings each day. The other half were asked to have eight servings of fruits and vegetables, plus 16 ounces of vegetable juice each day.

The researchers found that, indeed, the participants made significant diet changes.[30] In the eight-a-day group, fiber intake rose from 22 to 29 grams per day, and fat intake fell from 28 percent of calories to 21 percent within the first year. Estrogen levels fell, too. Serum estradiol concentrations fell from 91 pmol/L to 64 pmol/L. Estrone and estrone sulfate concentrations fell as well. That proves the point that changing your diet can indeed change your hormones. It really does work.

With blood tests, the researchers measured *carotenoids* in the women's blood. Carotenoids are *beta-carotene*—the orange pigment in carrots and sweet potatoes—and many related compounds. If a woman has a lot of carotenoids in her bloodstream, that proves she is eating vegetables and fruits. It turned out that those with the most carotenoids in their bloodstream had a 43 percent lower risk of either cancer recurrence or a new primary breast cancer, compared with women with lower blood levels of carotenoids.[31]

After seven years of follow-up, women in the five-a-day group who followed the guideline of eating at least five fruit and vegetable servings daily and who were also physically active turned out to have nearly a 50 percent reduction in mortality, compared with women who did not eat their fruits and vegetables or exercise.[32]

When the study results came out, it turned out that eight a day was not much better than five a day.[33] But the take-home message, from my viewpoint, is that vegetables and fruits matter, whether you get five a day, eight a day, or whatever generous amount works for you. So, it pays to not only cut out the animal products and fatty foods, but to also give vegetables and fruits—and exercise—a prominent role in your routine.

Fruit Favorites

When it comes to fruit, the whole group is beneficial, but some have emerged as superstars. In Harvard's Nurses' Health Study, researchers found that strawberries, blueberries, and peaches were strongly associated with reduced cancer risk.[34]

Soy Cuts Cancer Risk

As we have seen, traditional Japanese diets are associated with a remarkably low risk of developing cancer. Prominent in these diets are miso soup, tofu, and other soybean products. Could they get some of the credit for the low cancer rates?

Soybeans contain *isoflavones*, which have a chemical structure that is vaguely similar to testosterone or estrogen and, in test-tube experiments, have been shown to attach to estrogen receptors. At first, that led some to speculate that soy products might cause cancer. However, scientific studies have shown just the opposite. In numerous careful studies, women who consume the most soy products have been shown to have about 30 percent *lower risk of developing breast cancer*, compared with their soy-avoiding friends.[35,36] I should note that this has been demonstrated in Asian populations; in the United States, soy consumption is so low that reliable comparisons over the full range of high and low soy intakes have not been possible.

Similarly, in a 2012 study that included 9,514 breast cancer survivors, those who consumed the most soy products had a

30 percent reduction in cancer recurrence, compared with women who avoided soy.[37] An even larger study published the following year showed the same thing: Cancer survivors who included more soy products in their diets were much less likely to have a recurrence or to succumb to their cancer, compared with women who avoided soy products.[38]

It turns out that your cells have two different kinds of estrogen receptors: alpha receptors and beta receptors. Estrone, the main estrogen in women after menopause, binds mainly to the alpha receptor. Soy isoflavones preferentially bind to beta receptors, which appear to inhibit the cell proliferation that is part of the cancer process. Soy may also have other anticancer properties.

Many people—including some well-meaning but ill-informed physicians—have made the mistake of banning soy products for women with cancer, thinking they could increase recurrence risk. Research shows they have precisely the opposite effect.

For tackling cancer, soy is an ally. Soy products are not essential, of course. But to the extent they replace meat, milk, and so on, they are very helpful, and they appear to have a special ability to reduce cancer risk and boost survival.

Be Cautious with Alcohol

Alcohol increases breast cancer risk. Although many women unwind with a seemingly innocent daily glass of wine at the end of the day, it measurably increases risk, and there is no safe level of consumption.

Here are the numbers: If you are a young woman, each glass of wine in your daily routine boosts your chances of developing premenopausal breast cancer by about 7 percent, on average. Two glasses a day boost your risk by 14 percent, and so on. The same is true for other alcoholic beverages. A bottle of beer, a 5-ounce glass of wine, or a shot of liquor all have about the same alcohol content.* For postmenopausal breast cancer, the risks are worse. Every drink that you consume on a daily basis boosts your risk by about 13 percent. Double that for two drinks, and so on.[39]

What is alcohol doing? Alcohol contributes to DNA damage, and it may also have hormonal effects.[40] Needless to say, alcohol sometimes displaces healthier foods. In other words, if you are having wine, you are not pouring a glass of carrot juice.

Some evidence suggests that alcohol's effects are worse when your diet is low in folate, a B vitamin that plays a role in anticancer defenses. As its name suggests, it is found in foods with *foliage*: broccoli, Brussels sprouts, asparagus, and spinach. You will also find it in beans. It may be that including generous amounts of these foods in your routine will help protect you from alcohol's cancer-causing effects, at least to some degree. They will not eliminate it, however, so caution is still advised.

What About Sugar?

One might suspect that sugar could increase breast cancer risk. Sugar is a nutrient for the cells of the body, and that would presumably include cancer cells. So more sugar could mean more cell growth. And some tests have shown that when sugar is added to breast cancer cells in the test tube, they grow more rapidly than they otherwise do.[41] In addition, women with type 2 diabetes—who, after all, tend to have higher blood sugar values than other people—are more likely than others to develop breast cancer.

However, more recent studies have suggested that the association between diabetes and breast cancer is really just that both

---

* The actual numbers are a 9 percent increased risk for every 10 grams of alcohol consumed daily. A typical drink (12-ounce beer, 5-ounce glass of wine, or 1.5-ounce shot of liquor) has about 14 grams of alcohol.

tend to occur more frequently in people who have more body fat, rather than any sort of cause and effect.[42] The biggest problem with sugar may be that it lures us to consume doughnuts and candy, and these fatty, calorie-dense foods can lead to weight gain, which, in turn, fuels cancer risk.

So is it a good idea to kick a sugar habit? Well, yes, if you needed another reason to do that. But the key step for controlling blood sugar is to cut the animal products and added fats from your diet. This is because—surprising as it may sound—fat buildup inside muscle and liver cells causes *insulin resistance*, which keeps your blood sugar elevated. To put it simply, if your cells are full of fat, the cells cannot take in sugar very effectively, so it builds up in your bloodstream. Getting animal products and added oils out of your diet tackles insulin resistance and helps bring your blood sugar down. We'll have more to say about this in Chapter 8, "Conquering Diabetes." For now, the take-home message is that high blood sugar appears to fuel cancer, and shortly we'll look at how best to control it.

It Pays to Sweat

Exercise tames hormones. In the Women's Health Initiative, which we briefly discussed earlier, women who exercised had lower levels of estrogens circulating in their bloodstreams, compared with women who were sedentary.

There's a "however" here. It looks like much of exercise's hormone-taming effect comes from weight loss. Researchers at Fred Hutchinson Cancer Research Center in Seattle asked a group of women to exercise for forty-five minutes, five days a week.[43,44] Overall, their estrone, estradiol, testosterone, and other related hormones fell slightly, which is good. That ought to reduce their cancer risk. But the only women who benefited were those who lost weight during their exercise program. Those who did not lose any weight did not see their hormones drop at all.

Exercise has other benefits, though, apart from its effect on hormones. Studies suggest that it strengthens the immune system, helping you to eliminate cancer cells that may arise.[45]

Putting these elements together—a healthy plant-based diet, plus regular exercise—you can reduce your risk of breast cancer. And if you have been treated for cancer already, these steps will help you stay healthy.

Birth Control Pills, IUDs, and Hormone Replacement Therapy

What about the Pill? Does it cause cancer? Oral contraceptives contain hormones, typically estrogen and a progestin (a synthetic version of the hormone *progesterone*). As you might have guessed, oral contraceptives do raise the risk of breast cancer slightly. It had been hoped the newer brands containing small amounts of hormones would be less likely to cause cancer. Unfortunately, that does not appear to be the case. In a 2017 study that included 1.8 million women, newer formulations were linked to slightly increased cancer risk, too. Overall, for every 100,000 people, the use of oral contraceptives would be expected to cause 13 extra cases of breast cancer each year.[46] The good news is that this added risk quickly diminishes after use is discontinued, and studies suggest that oral contraceptives *reduce* the risk for cancer of the ovary, endometrium, and colorectum.[47]

Intrauterine devices (IUDs) are highly effective. Some brands contain a progestin; evidence so far suggests that their contribution to breast cancer risk is similar to that of oral contraceptives.[48,49]

Other IUDs contain copper, which is toxic to sperm cells. Some studies, although not all, show that copper-containing IUDs release a tiny amount of copper into the bloodstream.[50,51,52,53] Given that exposure to copper is one of the suspects in Alzheimer's disease,[54,55] this has raised the question as to whether long-term use of copper-containing IUDs could contribute to Alzheimer's risk. To my knowledge, no research has directly tackled this question.

To help women control hot flushes and other menopausal symptoms, doctors sometimes prescribe hormonal treatments, which take a wide variety of forms. These treatments have been studied for their contribution to cancer of the breast and other organs. We'll describe what researchers have found, as well as other approaches to menopausal symptoms, in Chapter 6, "Tackling Menopause."

Endometrial Cancer

Just as hormones boost breast cancer risk, they do the same for endometrial cancer. The endometrium is the inner lining of the uterus. During your reproductive years, the endometrium thickens every month and then, if pregnancy does not occur, is discarded in menstruation. The repeating waves of hormones make damage to DNA more likely to occur, and that can lead to cancer.

Endometrial cancer is more likely in women who have taken estrogens to control menopausal symptoms, unless they use a preparation that includes a progestin. Evidence suggests that birth control pills and vaginal estrogen creams do not increase the risk of endometrial cancer.[56]

So, what else can we do to protect ourselves from endometrial cancer?

Favor Foods That Trim Away Excess Weight

As we saw in the section on breast cancer, fat cells make estrogens, and the more estrogens in a woman's body, the higher her risk of endometrial cancer.[57] And a great way to reduce hormone excesses is to trim away excess body fat. In the Nurses' Health

Study, women at a healthy body weight had less than one-fourth the risk of endometrial cancer, compared with women who had a BMI of 35 or over.[58]

It's easier than you might think. In Chapter 12, we will look in detail at the best ways to trim down.

Choose Foods That Keep Your Blood Sugar Steady

Some studies have suggested that foods that stabilize blood sugar reduce your risk of endometrial cancer, while foods that boost blood sugar too much may elevate the risk for endometrial cancer.[59]

Simple substitutions will help: fruit instead of table sugar, rye or pumpernickel bread instead of wheat breads, sweet potatoes instead of white potatoes, and oatmeal instead of cold cereals. More on this in Chapter 8.

The most powerful way to reduce blood sugar actually has nothing to do with how much sugar or carbohydrate you eat. It turns out that when people begin a low-fat vegan diet, their blood sugars tend to fall in a healthful way. The reason is that this diet reduces the amount of fat inside the muscle and liver cells. In turn, these cells are then able to pull sugar out of the blood and use it as fuel. So, for a healthy blood sugar, a completely plant-based diet without added oils is the way to go. And then, for extra credit, you can use the substitutions described above.

Ovarian Cancer

Gilda Radner, a beloved comedian on NBC's *Saturday Night Live*, developed ovarian cancer in her late thirties. Despite treatments and a short-lived remission, the condition advanced aggressively. She died at age forty-two. Tragically, ovarian cancer has often already advanced by the time it is diagnosed.

Cancer cells arise most commonly in the fallopian tubes that lead from the ovary to the uterus. They can also start in the germ cells that will become eggs and the stromal cells that produce hormones.

What can you do about it? Unlike breast cancer, alcohol does not seem to play much of any role in ovarian cancer.[60] Tobacco does play a role, and quitting smoking helps.[61] Losing excess body fat reduces the risk, too, just as it does for many other cancers.

But an additional factor has emerged. In 1989, Dr. Daniel Cramer, whom we met in Chapter 1, examined ovarian cancer incidence in twenty-seven countries and found, surprisingly enough, that it paralleled milk consumption. The more milk women drank, the higher the incidence of ovarian cancer.[62]

Earlier, we discussed the fact that milk products have hormones, mainly because dairy cows are pregnant three-quarters of every year. Milk products are also typically high in fat, especially *saturated* ("bad") fat, which may also be linked to ovarian cancer.[63] But the reason for milk's apparent contribution to ovarian cancer risk may not be milk's hormones or fat. Rather, the issue appears to be the milk sugar—*lactose*. As we saw in Chapter 1, the breakdown products of lactose can be harmful to the ovaries.

In some populations, nearly everyone is lactose intolerant—that is, they no longer have the *lactase* enzymes that break lactose into glucose and galactose. These enzymes were in their intestinal tracts when they were breastfeeding babies but disappeared at some point after the age of weaning. This disappearance of *lactase* enzymes is the norm for blacks, Asians, Native Americans, and most everyone other than whites. Most whites carry a genetic mutation that causes the lactase enzymes to persist well into adulthood. While that may allow them to drink milk without digestive upset, it sets them up for more serious problems. Dr. Cramer reasoned that people who continue to be able to digest milk sugar would be exposed to galactose year after year. And if galactose can harm the ovary, then women could be at risk for ovarian cancer. And that is exactly what the evidence suggests.

Researchers at Rutgers Cancer Institute of New Jersey examined the diets of 490 women with ovarian cancer and compared them with women who had remained cancer-free, focusing on African American women, because they are at especially high risk. Those women who consumed the most milk had double the cancer risk, compared with those who consumed the least.[64]

Similarly, in Sweden, where dairy products are especially popular, a study following 61,084 women over thirteen years found much the same result. Those with the highest dairy consumption had twice the risk of ovarian cancer, compared with those who ate the least dairy products.[65] Other studies have confirmed that milk-drinking women have higher risk of ovarian cancer, compared with women who avoid cow's milk.[66]

Of course, Mother Nature never imagined that anyone would drink milk after the age of weaning, much less drink milk from another species. So hormones, galactose, and the other odds and ends in cow's milk should not be much of an issue. When people began to drink milk from cows and other animals, they exposed themselves to these compounds lifelong, something Mother Nature never had in mind.

Simple Dietary Steps for Tackling Cancer

To reduce cancer risk—or reduce your odds of recurrence—the following steps are in order. They will help you trim away unwanted pounds, avoid carcinogens, and tame excess hormones.

1. Avoid animal products. Throw out the dairy products, meat, and eggs. By focusing your meals on vegetables, fruits, whole grains, and legumes, you will avoid the hormones in

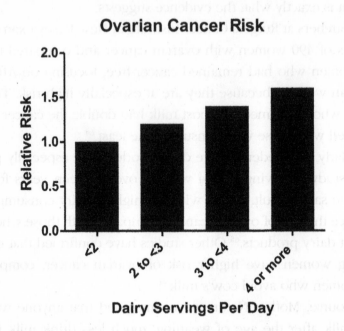

Ovarian Cancer Risk

Larsson SC, Bergkvist L, Wolk A. Milk and lactose intakes and ovarian cancer risk in the Swedish Mammography Cohort. *Am J Clin Nutr.* 2004;80(5):1353–1357.

dairy products and the carcinogens in meat, and will take advantage of the cancer-fighting nutrition of plant foods.

2. Minimize the use of added oils. While vegetable oils are far healthier than animal fats, they are still calorie-dense, refined products, and you will do better with non-oil cooking methods.

3. Eat for color, including orange and red fruits and vegetables and plenty of cruciferous vegetables (e.g., broccoli, cauliflower, cabbage, kale, collards, and Brussels sprouts). You'll get their antioxidants and other cancer-fighting nutrients.

4. Have tofu, tempeh, soymilk, or other soy products regularly.

5. Aim for 40 grams of fiber per day. Beans, vegetables, fruits, and grains will get you there easily.

6. Plan for two and a half to five hours of moderate-intensity physical activity every week. A brisk walk or anything else that gets your heart pumping is good.
7. When it comes to alcohol, the less, the better.
8. See Chapter 13 for tips on avoiding chemical exposures.
9. Be sure to take a vitamin B_{12} supplement. This is important for everyone, and especially those on a plant-based diet. You'll find more information in Chapter 12.

CHAPTER 4

Tackling Cancer for Men

Anthony Sattilaro was a physician and president of Philadelphia's Methodist Hospital. Hospitals are always busy, and with a new expansion project under way, Tony had his hands full. But it was exciting work, and he loved it.

One morning, he took a break to have a routine physical examination. There was not much to it. Apart from some aches and pains that he attributed to a minor bicycle accident, he felt fine. But as he arrived back at his office, the phone rang. The radiologist did not like the look of Tony's chest X-ray. He asked Tony to come back downstairs and have a look at it. Tony went down to radiology, and the radiologist put the X-ray on the screen. There was a large density in the left side of Tony's chest.

What the heck was this? Tony had had no chest symptoms at all. But this did not look good. The radiologist arranged for a bone scan the same day. And that was even more worrisome. The dense area on the X-ray turned out to be a cancerous tumor in one of Tony's ribs. The scan showed more tumors in his skull, sternum, and spine.

None of this was on the day's agenda. Suddenly, the busy physician was a patient with a very serious disease.

This was cancer, and a biopsy showed that it had started in his prostate, an organ just below the bladder. Cancer cells arise commonly in the prostates of older men. Often, the cancer advances slowly—sometimes so slowly it does not need to be treated. But Tony was only forty-six. To have advanced cancer at that age meant that he had an aggressive form of the disease. It had already spread throughout his body. Tony's doctors told him what he already knew. This was a death sentence. He had to get his affairs in order.

Tony tried to cope with it and to concentrate on his work as best he could. But soon the cancer became painful. As it ate into his bones, the pain gradually worsened, and eventually, he needed narcotic painkillers in order to function. And the painkillers caused nausea and other side effects.

To make everything worse, Tony's father had advanced lung cancer. Soon after his own diagnosis, his father died. When he got the news, he drove to New Jersey, where he buried his father and consoled his mother as best he could.

After the funeral, Tony headed back to Philadelphia. He was determined to throw himself into his work and do the best he could in the time he had left. As he drove onto the New Jersey Turnpike, he spotted two young hitchhikers and decided to pick them up, mostly to have someone to talk to. He told them about his work in the hospital, about the death of his father, and about his own diagnosis. He did not spare any details.

As fate would have it, the hitchhikers had just gotten out of macrobiotic cooking school. To them, cancer was no big deal. Why not change your diet and tackle this disease? It might well go away.

To Tony, these kids were completely irritating. With no medical knowledge whatsoever, they told him—an experienced doctor—that terminal cancer was curable. All he needed to do was to change his eating habits.

The word *macrobiotic* means "long life." Macrobiotic diets draw on the principles of traditional Chinese medicine and often incorporate traditional Asian foods, particularly brown rice, vegetables, miso soup, and other healthful foods, and they avoid dairy products, meats, and refined foods.

Tony let them drone on about yin and yang and how foods affect the body's energy. Blah, blah, blah. As he dropped them off, they asked for his address, promising to send more information. A few days later, a package arrived, sixty-seven cents postage due. Opening it up, he found a book about diet and cancer. He flipped through it, and it seemed like nonsense—the sort of faddism that preys on vulnerable people. But one thing got his attention. The book told the story of a woman—a physician—who had had breast cancer that had gone into remission after a diet change.

Hmmm. That's interesting, Tony thought. Breast cancer is a hormone-related cancer, just as prostate cancer is. He tracked down her phone number and gave her a call. Her husband answered. Tony asked how she was doing. Her husband said she was not doing well. She was dying of cancer.

Okay, case closed, Tony thought. "Thank you very much," he said. But the husband continued. Actually, she had done very well while on the diet—it really was miraculous for her. But she had found the diet hard to stick with and had gradually abandoned it. And now she was paying the price.

That got Tony's attention. Could there be something to this? Research on macrobiotic diets *per se* was pretty thin, Tony saw. But on the other hand, studies had suggested that healthful foods could cut cancer risk and maybe improve survival after diagnosis. And no one else had offered him any hope. He decided to give it a try. He went to Philadelphia's macrobiotic teaching center and asked for help.

Tony was all thumbs in the kitchen. He tried cooking brown rice but managed to blow up his pressure cooker. So he ate at the center and took extra food home. The tastes took a little getting used to, but he soon found foods he liked. At the hospital, he brought his brown rice and vegetables to the physicians' dining room, chopsticks and all, as the other doctors watched what seemed to be the desperate efforts of a dying man.

Except that Tony did not die. In fact, he began to feel better. His pain diminished day by day. Within three weeks, he was entirely pain-free. This was the first optimistic sign since his diagnosis. He began to feel normal again. As the weeks went by, he got his energy back and was able to concentrate on his work again. As the months rolled past, he found he was not thinking about his mortality.

A year later, his doctor repeated his bone scan. His doctor was stunned, and so was Tony. The cancer was nowhere to be seen. It looked like a totally normal scan. They had to assume that the cancer was not entirely gone, but it had shrunk to the point of being no longer visible.

This was unprecedented. His advanced cancer had disappeared. Tony decided that he had to find a way to let others know of his experience and of the possibility that foods are more powerful than doctors and patients had imagined. He wrote a book called *Recalled by Life*, which quickly became a bestseller, and he began to spread the word in lectures and television programs.

About a decade later—long after Tony was supposed to have succumbed to cancer—he was still doing well. I heard about his amazing recovery and gave him a call. He invited me to visit him in his home, which I did. He was fit and healthy. He showed me his scans, and the before-and-after evidence was indeed astounding.

Regarding the effect of diet, he remained cautious. The diet change had done great things for him, but he was not sure what it would do for other people or for other kinds of cancer.

But then Tony told me something troubling. He had been well for so long, he had decided to test himself. He wanted to see if he would still be well if he went off the macrobiotic diet for a while.

I have to say, this made me nervous. His macrobiotic counselors had strongly urged him not to rock the boat. Driving cancer into remission once is a challenge. Taking a chance that it would return and then trying to tackle it a second time was something no one other than Tony wanted to try.

Some months later, I called Tony. His voice was different. His speech was slurred, as if he was on narcotics. And, yes, he told me, the cancer had returned. He was in terrible pain. And a few weeks later, Tony died.

Did an unhealthy diet cause his cancer? Did the diet change make it go away? If he had stuck with a healthy diet, would he still be alive today? There are no answers to these questions. But many other people have also used a healthy diet to fight cancer, and research studies are starting to catch up with them, showing the power of foods.

Of course, this does not mean that surgery, radiation, chemotherapy, and other cancer treatments do not have important roles. They do. But, in addition to whatever other treatments a person may need, food choices are key, often most important of all.

Using Foods to Reduce Prostate Cancer Risk

So what do we know about foods and prostate cancer, and how can men protect themselves?

First, body weight plays a role, just as it does in many forms of cancer. For overweight men, the problem may be insulin resistance. Normally, insulin escorts glucose and protein into cells. But accumulating body fat causes cells to resist insulin's action. In response, the pancreas makes more and more insulin, trying to overcome this resistance, and this extra insulin appears to increase cancer risk.[1]

Slimming down helps. With less body fat, insulin can do its job without going overboard. You'll find more details in Chapter 8, "Conquering Diabetes."

As with all hormonal conditions, a plant-based diet has special power. Over the next several pages, we will look at four specific parts of the diet that you should be thinking about for cutting your risk of prostate cancer. Two appear to increase risk, and two others reduce it.

Links with Milk

A surprising body of evidence has linked dairy products with prostate cancer. For starters, the disease has long been common in the United States and Northern Europe, where milk, cheese, and butter are the order of the day.[2] It was rarer in Japan, Thailand, China, and other countries where dairy products were not part of the traditional diet. As Westernization brought dairy and meat products into these countries, prostate cancer rates climbed.[3]

Two large Harvard studies investigated the connection and found that milk-drinking men are much more likely to develop prostate cancer, compared with men who generally steer clear of dairy products. The first of these, called the Physicians' Health Study, included 20,885 men and found that those who had at least two and a half dairy servings per day had a 34 percent increased risk of developing prostate cancer.[4] The second, called the Health Professionals Follow-up Study, included 47,781 men and found that men drinking more than two milk servings per day were 60 percent more likely to develop prostate cancer.[5]

In 2016, researchers looked at the full body of evidence to date. Combining the results of eleven different studies, they found that men consuming the most milk products had a 43 percent higher risk of dying of prostate cancer, compared with men who generally avoided dairy products.[6]

Why would this happen? Does milk have hormones in it? Yes, of course it does. It comes from a cow who is busily making hormones—especially estrogens—that get into her milk and become concentrated as milk is turned into cheese. But for prostate cancer, the real problem might be something else. When you drink milk, your body produces a compound called *insulin-like growth factor* (IGF-1). Everyone has some IGF-1 in their blood naturally, especially during childhood and adolescence. As its name suggests, it helps you grow. When you have reached adulthood, your body makes less IGF-1, because it's time for growth to stop. In fact, you do not want too much IGF-1 in your blood as an adult, because, just as it causes children to grow into adults, it also causes cancer cells to grow into tumors. If you were to mix IGF-1 with cancer cells in a test tube, the cells would grow like crazy.

Milk consumption boosts IGF-1 levels. In a Creighton University study, adults who had three dairy servings a day had increases in their IGF-1 levels of about 10 percent.[7] So, could a 10 percent rise in IGF-1 boost your cancer risk? Well, let's go back to Harvard University and look again at the Physicians' Health Study. The researchers checked blood tests on each participant. It turned out that those men who developed cancer had had about 10 percent more IGF-1 in their blood when the study started, compared with men who remained cancer-free.[8] In other words, it appears that dairy products boost the production of IGF-1 in the human body and, in turn, IGF-1 boosts cancer cell growth.

Yikes! That's a pretty bad advertisement for the dairy case.

And there is more. Dairy products are linked to prostate cancer through another mechanism, this one relating to vitamin D. Normally, vitamin D comes from sunlight. As sunlight hits your skin, it produces a form of vitamin D that is then activated in the liver and kidneys. From there, vitamin D works as a hormone whose job is to help you absorb calcium from the foods you eat. Vitamin D also has a cancer-preventive effect.

And here is where dairy products do some dirty work. If you drink milk or eat cheese, their calcium floods into the bloodstream. Your body recognizes that it is actually getting *too much* calcium and it starts to slow down the activation of vitamin D, aiming to reduce calcium absorption. That's a good reaction. It will keep your body from overdosing on calcium. But when you turn off your vitamin D activation, you lose part of its anticancer effect. So dairy products have the effect of *suppressing* your body's vitamin D activity, and that puts you at higher risk for cancer, or at least that is what evidence suggests.

While scientists continue to tease out more details about how milk products cause cancer, it is useful to remember that nature "designed" milk for only one purpose—to provide temporary nourishment for a rapidly growing infant. If we defeat the weaning process and continue to make drinking milk a lifelong habit, we make ourselves vulnerable to effects that nature never intended.

Fish Oils and Prostate Cancer

Certain omega-3 fatty acids have also been implicated in prostate cancer. They are the long-chain omega-3s found in fish oils: eicosapentaenoic acid (EPA), docosapentaenoic acid (DPA), and docosahexaenoic acid (DHA).* For years, supplement manufacturers have been pushing fish oils for heart health, brain health, and many other purposes. Most of these health claims have

---

* If you could look at an omega-3 molecule under a powerful microscope, it would look like a chain of carbon atoms joined together. The name *omega-3* comes from the fact that the first double bond in the chain comes after carbon atom number three, counting from the chain's end (omega = end).

Plants make an omega-3, called *alpha-linolenic acid* (ALA), whose molecular chain is 18 carbons long. Your body can lengthen it to chains of 20 carbons (EPA) or 22 carbons (DHA), and it is the DHA that may be important for brain health.

The enzymes that lengthen the 18-carbon chain to 20 and 22 carbons are not very efficient and can be slowed down if your diet has lots of other fats that compete with omega-3s for their attention.

crashed and burned under scientific scrutiny. In the process, several research studies have found that men with higher amounts of these omega-3s in their blood had higher prostate cancer risk.

The question was, is this real? Does fish oil cause prostate cancer? Or is it an innocent bystander? There was no obvious biological mechanism by which fish oil should cause cancer. Perhaps health-conscious men just tend to choose fish or fish oil supplements and also get checked for prostate cancer more often than other men, leading to more frequent cancer diagnoses. A 2007 Harvard study did not find the link and, in fact, showed the opposite.[9] Maybe it was all a fluke.

But in 2013, Ohio State University researchers reported new findings from a large group of men. And it turned out that those with the most EPA, DPA, and DHA in their bloodstreams had 43 percent higher prostate cancer risk, compared with men with the lowest blood levels. They wrote, "Whereas a lack of coherent mechanism has led authors of previous studies, including us, to consider these findings suspect, their replication here strongly suggests that long-chain ω-3 PUFA [that is, EPA, DPA, and DHA] do play a role in enhancing prostate tumorigenesis."[10] In other words, believe it.

In 2014, researchers from seven different studies put all their data together. They had findings from 5,098 men who developed cancer and 6,649 who did not. It turned out EPA and DHA were both linked to prostate cancer. Men with the highest EPA blood levels had a 14 percent higher risk, compared with men with the lowest levels. Men with the highest DHA blood levels had a 16 percent higher risk.[11]

So what do we make of this? First, plant-derived omega-3, called *alpha-linolenic acid* (ALA), does not appear to cause prostate cancer. It is actually an essential nutrient, whether it comes from walnuts, flaxseeds, or green vegetables.

For EPA, DPA, and DHA, we simply do not know whether they increase cancer risk or if these findings are just an artifact of the research design and of the increasing cancer detection caused by the emergence of more routine cancer screening. This is an important question, because some people choose to supplement with EPA and DHA, hoping to reduce their risk of Alzheimer's disease. This is a reasonable decision, although evidence of benefit is not yet in.

If you do choose to take EPA and DHA, I would recommend the vegan brands (derived from algae), which are widely available online. They help you avoid the impurities and odors that can come from fish-derived versions. However, there is no reason to think their health effects—good or bad—are different from fish-derived EPA and DHA.

Tomatoes to the Rescue

The red color in a tomato, watermelon, or pink grapefruit comes from lycopene. It is a cousin of beta-carotene, the pigment that makes carrots orange. Both are powerful antioxidants, meaning they can neutralize the free radicals that could otherwise cause cancer. A Harvard study found that men who have ten or more servings of tomato products each week have 35 percent less risk of developing prostate cancer, compared with men who have tomatoes less often. And that was true even when the tomatoes came in the form of ketchup and pizza sauce.[12]

In 2017, researchers at the University of Illinois combined the results of forty-two prior studies and confirmed the finding: The more lycopene you eat, the lower your prostate cancer risk.[13] Raw tomatoes are fine. But cooking helps release lycopene from the tomatoes, making it easier to absorb.

Another Advantage of Soy Products

Like other hormone-related cancers, prostate cancer is less common in Asian countries, compared with North America, and research has suggested that soy products may be one reason why.

Of course, part of the value of soy products comes from what they are not. That is, soymilk is not cow's milk, and a soybean burger is not meat. Given what we know about milk and meat, those are real health advantages. But soybeans also have protective hormonal effects, as we saw in the previous chapter.

In 2018, researchers combined the results of prior studies, finding that men consuming the most soy had 29 percent less risk of developing prostate cancer, compared with their soy-avoiding friends. They then asked whether unfermented soy products (tofu, soymilk, and edamame) had an effect different from fermented products, such as miso (a soybean paste used in soups) and natto (fermented soybeans with a pungent taste and smell). The winners were the unfermented foods, which cut cancer risk by 35 percent.[14]

So, to reduce your risk of developing prostate cancer, there is value in avoiding dairy products (and animal-derived products in general) and taking advantage of the power of plants, especially soy foods and brightly colored, antioxidant-rich foods, like tomatoes and watermelon. For omega-3s, the jury is still out.

If you have been diagnosed with prostate cancer, similar dietary changes can help you, too, as we will see in the next section.

Prostate Cancer Survival

As Anthony Sattilaro's case illustrates, foods can influence not just whether cancer starts but whether it advances or regresses. The foods that helped Tony were plant-based. This parallels a finding made by researchers from the California Cancer Registry, who found that when Japanese American men were diagnosed with prostate cancer, they were less likely to die of it—34 percent less likely, to be exact—compared with white Americans.[15] Needless to say, a traditional Japanese dietary pattern is very different from a typical American diet. It has no milk at all and not much meat.

It emphasizes rice, vegetables, fruit, and soy products, such as miso soup and tofu. It is not exactly a macrobiotic diet, but it is much closer to it than the meaty meals most Americans grow up on. Although Japanese traditions have eroded considerably under Western influences, to the extent remnants of those traditions remain, they have made for healthier diets.

The effect of foods on prostate cancer was put to the test by Dr. Dean Ornish, who was already famous for showing that a vegetarian diet, along with a healthy lifestyle, could reverse heart disease. In 2005, he published a new study that included ninety-three men with prostate cancer, divided into two groups.[16] In one group, men were asked to follow their usual eating habits; that was the control group. The other group was asked to do something very different. They began a diet with no animal products at all—that is, a vegan diet. So, out with the chicken, fish, turkey, and burgers, and in with healthy grains, vegetables, fruits, beans, and the various meals that are made from them. If a man wanted bacon or sausage, it had to be veggie bacon or veggie sausage. Spaghetti was to be topped with tomato sauce instead of meat sauce.

The research team followed the men's health with a test called *prostate-specific antigen* (PSA). A rising PSA level indicates increased activity of prostate cells. For men with prostate cancer, it suggests cancer progression.

For the men following their usual diets, the average PSA level gradually worsened, as it typically does in prostate cancer. Over a year's time, it rose 6 percent. And of the forty-nine men in that group, six could not continue in the study—their cancer was advancing too quickly, and they needed surgery or radiation.

For the men on the vegan diet, the experience was just the opposite. Their average PSA wasn't rising. It actually *fell* about 4 percent, which was good news. And not one of the men on the vegan diet needed treatment during the year-long study.

After two years, thirteen of the forty-nine men in the control group had had to undergo surgery, radiation, or other treatment for their cancer, compared with only two of the forty-three men on the vegan diet.[17]

These studies show that prostate cancer is strongly linked to Western dietary habits. The hormone haywire caused by these foods contributes to the proliferation of cancer cells. Changing to a plant-based diet makes cancer less likely to occur and improves survival for men who have cancer already.

Testicular Cancer

Testicular cancer is the most common cancer in men between the ages of twenty and forty-five. It is often detected early and is easily treated. As a result, survival rates are high. However, it is increasingly common, suggesting that something other than genes and bad luck is at work.

The problem does not appear to be smoking or alcohol. Neither one has shown any relationship with testicular cancer. But hormones do play a role. Exposure to estrogens during uterine development appears to increase risk. And, like prostate cancer, testicular cancer is more common in men with a higher body weight.

Dairy products, especially cheese, have been identified as suspects in several studies.[18] A 2003 study showed that men consuming the most cheese had 87 percent higher risk of developing testicular cancer, compared with men who ate little or no cheese.[19] Needless to say, dairy products contain estrogen traces that are more concentrated as milk is turned into cheese.

Processed meats (e.g., sausage, bacon, turkey bacon, ham, and hot dogs) are also linked to higher risk. You probably did not need another reason to avoid these foods, which are well-known contributors to colorectal cancer and many other health problems.

Steps for Tackling Cancer Risk

The steps for reducing your risk of prostate and testicular cancer and for improving survival are similar to those for other cancers. Our goal is to trim away unwanted pounds, avoid carcinogens, take advantage of cancer-fighting foods, and tame excess hormones.

1. Avoid animal products. Throw out the dairy products, meat, and eggs. Focus your meals on vegetables, fruits, whole grains, and legumes. These foods help weight control and bring you cancer-fighting nutrition.
2. Minimize the use of added oils. While vegetable oils are healthier than animal fats, avoiding added oils will accelerate healthful weight loss.
3. Take advantage of lycopene-rich foods, such as tomatoes and watermelon.
4. Have soy products regularly: soymilk, tofu, tempeh, edamame, miso, and so on.
5. Go for 40 grams of fiber per day. Beans, vegetables, fruits, and grains will get you there easily.
6. Be sure to take a vitamin B_{12} supplement. This is important for everyone, and especially those on a plant-based diet.

CHAPTER 5

Reversing Polycystic Ovary Syndrome

Polycystic ovary syndrome (PCOS) can be a challenge for women, and for their doctors, too. The symptoms are often confusing, and so is the name, since you can have PCOS even without cysts on your ovaries. Let's make sense of this condition and what you can do about it.

Alison

Alison is an oncology dietitian, practicing in Wisconsin. Many have sought her help in dealing with cancer and its complications, and she has found that diet changes can be enormously helpful. But Alison had a problem of her own.

Generally, her health was great. She was at a healthy weight, athletic, and full of energy. But something was not right with her monthly cycle. "From the moment I hit puberty and got my first period, I was never regular," she said. "I would go for five months without a period, and when it did arrive, it was very heavy."

In college, she noticed some dark facial hairs. Some other women in her family had the same thing. She assumed it was just her Greek ancestry. She also developed acne.

At age twenty-three, she married Patrick, whom she had been dating for several years. They looked forward to raising a family. Three years later, they were out of school and settled, and the time was right.

However, her cycle would not cooperate. She was not ovulating. In a year's time, she did not have more than one or two periods. So she carefully tracked her body temperature and other parameters noted to be important in identifying a fertility window. Her doctor looked at her numbers and could not make heads or tails of them. There was no trend at all.

Eventually, her doctor diagnosed PCOS, a hormonal condition manifesting as irregular periods, acne, and changes in hair growth—either unwanted body hair or hair thinning. The name comes from ovarian cysts, which are large follicles surrounding the eggs; some (although not all) affected women have these. PCOS is a common cause of infertility.

"I wasn't surprised by the diagnosis," she said. "Other close family members had been diagnosed before me, and I had the typical symptoms."

Diagnosing the problem was one thing. Fixing it was another. Without a regular cycle, pregnancy was not likely. Her doctor prescribed metformin, a medicine used to treat type 2 diabetes that is also used for PCOS. It did not improve her fertility at all.

A year passed, and pregnancy tests stayed negative. Her doctor prescribed fertility medications. They did not seem to work, either.

Through it all, Patrick was completely supportive, she recalls. "But I felt as though it was all my fault," she said. "It was my body that wasn't cooperating. I foolishly would ask him, 'If I can't give you children, will you still love me?' He held me. He let me cry."

All the while, she was prescribing healthy diets to help her cancer patients. And suddenly, it hit her: Maybe a diet overhaul would help her, too. She stopped eating meat and started weaning herself off dairy products. One day, she decided to throw out animal products altogether, to keep sugars and processed foods to a minimum, and to follow as healthful a diet as she could. "And three weeks later—to the day—I ovulated for the first time in a year. And about three weeks after that, I learned that I was pregnant." She had a healthy pregnancy and gave birth to a beautiful seven-pound, nine-ounce baby girl.

"I'm a Wisconsin girl. If you had told me that I would one day follow a vegan diet, I would have thought you were crazy." She noticed several other changes, too. "My acne is almost nonexistent, my GI troubles have drastically improved, and I lost the few pounds I had been wanting to lose but previously couldn't—even post-pregnancy."

She created Wholesome LLC, her private practice dedicated to giving patients and their caregivers the information and inspiration they need to put the power of food to work for themselves. As a family of three, Alison and Patrick continue a whole-food, plant-based diet and make burgers, tacos, chili, and spaghetti without meat and dairy products. They continue to explore new foods, new tastes, and a new level of health as they watch their baby girl grow, explore, and develop—something they always dreamed of and now live out.

Understanding PCOS

In 1935, two American gynecologists, Irving F. Stein and Michael L. Leventhal, described a condition in which women had fluid-filled cysts in their ovaries and ovulation patterns that were off track.[1] Affected women often had irregular periods and other problems.

Over time, however, it became clear that cysts were not the main issue. Rather, they were a symptom of a deeper problem. It turned out that the real issue in PCOS is excess androgens ("male hormones").[2,3]

It is perfectly normal for your ovaries to produce traces of androgens. They do not have much obvious effect, because most are converted to estrogens ("female hormones"). But in PCOS, the ovaries produce extra androgens, and the adrenal glands, situated just above the kidneys, do the same thing. You end up with too much in the way of male hormones, and they can cause problems:

1. Irregular periods, or going for months without any period at all. Sometimes the condition leads to heavy or prolonged periods.
2. Skin changes: acne, unwanted facial or body hair, and male-pattern hair loss.
3. Polycystic ovaries, meaning enlarged ovaries with follicles surrounding the eggs.

As we have seen, PCOS interferes with fertility. And when women with PCOS do become pregnant, they are at a higher risk of developing gestational diabetes and high blood pressure. The good news, as we will see below, is that well-chosen foods can help you tackle all these problems.

For some women, PCOS symptoms are barely detectable; for others, they are more pronounced. And not everyone gets every symptom. As I noted earlier, "PCOS" is a misnomer, because some affected women do not actually have polycystic ovaries. What they do have is extra androgens.

Androgens cause other problems. They change your body fat pattern so that it tends to end up at your waistline, rather than on your hips and thighs. This did not happen to Alison, but it does happen frequently in PCOS. And this belly fat works more mischief. It encourages even more androgen production in the adrenal glands and the ovaries.

Many women with PCOS develop *insulin resistance*, meaning that their cells do not respond well to their bodies' natural insulin. That leads to rising blood sugars and sometimes diabetes. Over the long run, other problems can arise: infertility, depression, heart problems, and a higher-than-usual risk of endometrial cancer.

So what can we do about it? The condition is partly genetic. Even so, there are several ways to treat PCOS. Doctors prescribe antiandrogens and oral contraceptives to counter the effects of androgens, and they prescribe metformin to reduce insulin resistance. These medications have their places. But foods play key roles. Properly chosen, foods can trim away excess weight, eliminate hormone excesses, and counter insulin resistance to reduce blood sugar. Let's look at each of these.

First, losing weight. Many—although by no means all—women with PCOS are carrying unwanted weight. With weight loss, PCOS can markedly improve or even disappear. The key is to go for *healthy* weight loss—not a crash diet—and to make it stick.

As you will see below, we will not focus on *how much* you eat. Rather, we'll focus on *what* you eat. We'll build our meals from plant-based foods and take advantage of cooking methods that minimize added oils. These meals work magic in two ways: They satisfy your appetite before you have overdone it with calories, and they give your metabolism a natural boost so you burn calories faster. This combination of appetite control and a better metabolism will help you trim away unwanted weight and keep it off long-term.

Second, controlling hormones. Of course, losing weight alone will help you control hormones. As you know by now, body fat is a hormone-building factory, and getting rid of excess body fat with a plant-based diet will help you get into better balance.

Foods from plant sources have an extra hormone-taming action, apart from their effect on your waistline. As you will recall from Chapter 1, "Foods for Fertility," this is due to their low fat content and to fiber's natural ability to escort unwanted hormones out of your body. Your liver removes hormones from the bloodstream and sends them into the intestinal tract. If you have plenty of fiber in your diet, it traps those hormones in the intestinal tract and carries them out with the wastes.

Third, controlling blood sugar. PCOS is different from the other hormone-related conditions we have discussed so far in that the issues go beyond sex hormones. As weight creeps upward, blood sugar control often gets off-kilter. Over the long run, that spells a higher risk of diabetes, heart disease, and other health problems, and can even aggravate androgen excesses. Getting your blood sugar under control will cut your risk of all these problems. Let's take a little time and go over what you need to know:

Glucose, a simple sugar, is your body's main fuel. Just as gasoline powers your car, glucose powers your body. All parts of your body, including your muscles and your brain, run on glucose. So while some people think of sugar as something bad or indulgent, it is actually what keeps your body moving. It is essential to life. Starches and sugars that you eat give you glucose, and that's all good.

To get gasoline into your tank, gas stations have fuel pumps. To get glucose into your muscle or liver cells, your body has *insulin*, a hormone made in your pancreas. Arriving at the surface of each cell, it opens up the cell membrane to allow glucose to enter.

Here is the problem: If fat builds up inside your muscle cells, it can interfere with insulin. Let's say, for example, you have bacon and eggs for breakfast. Fat particles from these foods pass into your muscle cells. If your lunch is chicken salad and your dinner is cheese pizza, these fatty foods pack more fat into your cells. If you keep this up day after day, that fat can eventually stop insulin from working normally. If insulin can no longer open the cell membrane to the passage of glucose, glucose builds up in the bloodstream. As your blood glucose level rises, it leads to prediabetes and eventually diabetes.

By the way, we are not speaking of *body fat* for the moment. I'm referring to tiny particles of fat *inside your muscle and liver cells* that are causing insulin to misfire.

Your body tries to compensate for insulin resistance by making *extra* insulin. If the body can make enough insulin, sooner or later glucose will be driven into the cells. But that excess insulin causes problems. It reduces the amount of SHBG in your blood. As you'll recall, SHBG is that group of "aircraft carriers" that hold on to estrogens and androgens, keeping them temporarily inactive. If you have less SHBG, your androgens become more active—exactly what you *don't* want.

Our research team found that food choices can help you reduce the amount of fat inside your muscle and liver cells, allowing you to counteract insulin resistance and restore *insulin sensitivity*.

Food Choices for Tackling PCOS

You want to lose weight, rein in your hormones, and control your blood sugar. Here are the steps that can help you accomplish each of these goals.

1. Avoid animal products. The most powerful eating plan eliminates *all* animal products—fish, chicken, dairy products, eggs, and so on—and relies instead on vegetables, fruits, whole grains, and beans. These foods are naturally low in fat and bring you abundant fiber.

Start your day with a bowl of old-fashioned oats topped with cinnamon, raisins, and sliced strawberries, or whole-grain pancakes with blueberries, maple syrup, and veggie sausage. Lunch could be a chunky vegetable soup, savory bean chili, sweet potato curry, or a veggie burger with lettuce, tomato, onions, and pickles. Dinner might start with a piping-hot bowl of minestrone or lentil soup and a green salad, followed by angel hair pasta topped with a sauce of tomatoes and wild mushrooms, and finish with fresh strawberries and an espresso, if you like. Or have a spinach enchilada or bean burrito. If your friends are getting together at a sushi bar, have a cucumber roll or sweet potato roll and a seaweed salad. The idea is to take advantage of the power of plant-based foods. These foods have no animal fat, obviously, and are very low in fat overall. And they are loaded with healthful fiber.

What a Little Soy Can Do

Canadian researchers found that natural compounds in soybeans and their cousins—chickpeas, peas, and lentils—seem to work at the cellular level to reduce insulin resistance and block body fat storage.[4] So, could having some soymilk or tofu help women with PCOS?

In a 2016 experiment, researchers randomly asked women with PCOS to take 50 mg of soy isoflavones (natural compounds in soy products) each day. That is the equivalent of 2 cups of soymilk or 7 ounces of tofu. After twelve weeks, the amount of androgens in their bloodstreams had diminished, as had insulin resistance.[5]

In a later experiment, researchers asked thirty women with PCOS to replace half of the animal protein in their diets with soy protein, while another thirty women continued their usual diets. The diet change led to healthy drops in weight and blood sugar and also reduced the amount of androgens (testosterone) in their bloodstreams.[6]

Of course, part of the value of soy products is that they kick meat and dairy products off your plate. But they and their botanical relatives appear to have benefits of their own.

---

2. Avoid fatty foods and added oils. Even though vegetable oils are far safer than animal fats, evidence suggests that they can contribute to insulin resistance, too, not to mention weight gain. So it pays to keep oils to a minimum, too. Chapter 12, "A Healthy Diet," has plenty of tips that will make it easy. You'll find that there are plenty of ways to cook without adding oils. Soon, you will come to prefer the lighter taste and super-quick cleanup. It also helps to minimize the use of fatty foods, such as nuts, peanut butter, and avocados.

3. Favor low–Glycemic Index foods. The Glycemic Index is a simple rating system that shows which foods are gentlest on your blood sugar. For example, white bread raises blood sugar quickly. But rye bread is gentler on your blood sugar, and pumpernickel even more so. They do not affect your blood sugar the way that white bread does. So white bread is a "high–Glycemic Index" food—or high-GI food, for short, and rye and pumpernickel are "low-GI." You want to favor low-GI foods.

With some simple substitutions, you can shift from high-GI foods to lower-GI foods. Have a look at Chapter 8, "Conquering Diabetes," for details.

By avoiding animal products, minimizing oils, and favoring low-GI foods, the fat inside your cells will gradually dissipate. In the process, insulin resistance fades away. This helps bring down high blood sugar, and it improves your overall body chemistry and hormone balance.

It turns out that as your insulin level falls back toward normal, your body's production of SHBG rises, helping you tackle hormone excesses.[7]

Don't Cut Carbs

Earlier, I suggested being selective about your carbohydrate-containing foods, choosing those with a low GI. Note that I did not say to avoid carbohydrates. In the past, people used to imagine that a good way to bring down high blood sugar was to avoid starchy foods. That was an understandable idea, but not a very helpful one. Avoiding carbohydrates does not fix insulin resistance, and that's the real problem.

The populations that are the slimmest, healthiest, and live longest have not followed anything like a low-carb diet. Quite the reverse; they have eaten an abundance of carbohydrate-rich foods: rice in Japan, sweet potatoes in Okinawa, and pasta in southern Italy, for example. Low-carbohydrate fad diets (Atkins, South Beach, ketogenic diets) cause weight loss only when you omit enough foods that calorie intake falls. In the process, the meaty, greasy foods that are emphasized in these diets often lead to cholesterol problems and increased mortality over the long run.

One other point: While part of the rationale for low-carbohydrate diets is based on the fact that carbohydrates stimulate insulin release, it is important to know that protein triggers insulin release, too. Researchers at the University of Sydney carefully calculated the insulin release caused by a wide variety of foods, finding many surprises: Fish and beef stimulate insulin release to a greater degree than pasta, popcorn, or peanuts.[8]

Most importantly, since we aim to tackle insulin resistance, we want to use a diet that removes fat from the cells most effectively, and that is a low-fat, plant-based diet.

Tackling PCOS means trimming away weight in a healthful way, and using foods to tackle androgen excesses and insulin resistance. It pays to draw your nutrition from plant sources, keep fats low, and choose healthful, low-GI foods that keep their fiber intact. As you will see in the recipe section, these guidelines translate into an abundance of healthful and delicious meals.

CHAPTER 6

Tackling Menopause

Menopause is a part of life. Even so, it can bring its share of surprises: hot flushes, unpredictable emotions, changes in sexual function, and memory fog. Most women sail through menopause with no big problems. But some are hit harder by the experience. When symptoms occur, the cause is shifting hormones. In this chapter, we'll tackle them.

First, a note of reassurance. If you have found yourself on a roller coaster of hot flushes and emotional changes, all of this is likely to pass on its own. It may take time, but you will eventually feel much more in control of your life.

Ann

Ann was fifty, living in Portland, Oregon, when it started. All of a sudden, she became uncomfortably hot and broke out in a sweat. After a few minutes, it stopped. It happened again and again; soon she learned that hot flushes have a mind of their own.

Hot flushes are uncomfortable for anyone. But under her bullet-proof vest and gun belt, they took on a whole other dimension. As a patrol officer on the Portland police force, she could not just flip up her shirt and fan herself. She had to deal with them.

Hot flushes came at night, too. "I'd wake up two or three times every night, and I'd have to throw the covers off, and I kept a fan by my bedside," she said. "But if you don't sleep well, you're fatigued the next day. As a police officer, you need to be alert."

Before long, she found a possible solution. She had been raised in Arkansas on beef, chicken, pork, and occasional vegetables. She loved milk and had plenty of cheese and butter in her routine. And she had a terrific craving for chocolate. But research studies had shown that soy products were effective against hot flushes. As we will see below, they work for some women and not for others, but sometimes they give substantial relief.

"So I picked up a can of soy powder and stirred it into soy-milk and drank it in the morning before work. And within a couple of weeks, the hot flushes were much less frequent and less intense. They dropped from several times a day to just a few times a week."

Sometimes she was too rushed to mix up her soy drink. "If I skipped the soy powder drink for a few days, the hot flushes returned. When I resumed the drink, they disappeared again. That convinced me that this was cause and effect. As long as I had that daily drink, my hot flushes were much more manageable."

Marie

Marie grew up on a dairy farm in Tennessee, the younger of two girls. Her health was excellent, apart from some joint pains.

She was fifty years old when, out of the blue, she had an unusual warm sensation. What was this? A fever? Maybe she was coming down with something. After a few minutes, it passed. But before long, the warmth came back and was more intense. As it recurred, she realized that, of course, this was menopause.

Unfortunately, the hot flushes worsened and eventually became debilitating. "I had them ten to twelve times a day," she said. "In public, it was embarrassing. My whole upper body turned beet red, and it was like steam was coming out of my ears. I'd be fine for a couple of hours, and then it would come again." Every night, she was awakened a half-dozen times and could never get a good night's rest.

After two years of this, it was clear that her hot flushes were not going away. Her doctor recommended hormone replacement therapy (HRT). HRT was not just a way to control hot flushes; it promised to delay wrinkles and help women stay younger—or so doctors said at the time. She agreed to try it and found that, indeed, HRT eased her symptoms.

Eventually, however, she was concerned about the risks of continued hormone treatments and talked with her doctor about stopping them. He agreed. But as soon as she quit HRT, hot flushes returned with a vengeance.

"I felt I would spontaneously combust," she said. "With intense hot flushes throughout the day and waking up five or six times every night, I was about ready to kill myself or kill somebody else. I'm joking, but I really felt terrible." After several weeks, she had had enough and resumed HRT. Twice more, she tried to stop the treatments, to no avail. Three months was the longest she could go before the symptoms were intolerable.

When she was sixty-four, however, her thirty-seven-year-old daughter was diagnosed with breast cancer (happily, she is doing well now). Her daughter insisted that Marie get off HRT, which she saw as an invitation to breast cancer. So Marie did quit, this time for good. The hot flushes came back again, as bad as ever, and she just put up with them.

Three years later, she decided to tune up her diet. She eliminated meats, dairy products, and processed foods, and stopped adding oils to foods. She ate a lot more fruits and vegetables. And good things started to happen. After about three months, it hit her. "I'm not waking up in the middle of the night anymore. I'm not having terrible hot flushes during the day."

Her hot flushes were not gone, but they were mild and infrequent. "I may have one or two a day, and I just feel a degree or two warmer, not really hot. At night, I am occasionally warm, but I'm not burning up. My family no longer has to joke about finding me as a heap of ashes."

In the bargain, she lost thirty-five pounds. Long-standing joint pain subsided, as did sinus problems, and her energy level skyrocketed. "I can go out and run five miles continuously, and I feel great," she said.

Today, she prefers simple foods. Breakfast might start with steamed greens, followed by oatmeal with berries or a banana. Lunch or dinner could be a salad, beans and rice with salsa, or baked sweet potatoes, and generous amounts of vegetables. "I never thought I liked vegetables before," she said. "Now I love roasted squash, roasted Brussels sprouts, and steamed cabbage. I center my meals around vegetables." Dessert is grapes, apples, pears, or other fresh fruit—whatever is in season.

What Is Menopause?

During the reproductive years, the release of an egg each month is followed by a surge in the ovary's production of estradiol and progesterone. But at menopause, the ovaries are running out of viable *oocytes*, or eggs. With ovulation no longer happening, the amount of estradiol and progesterone drops sharply. And that leads to many changes in your body. Doctors use the term *menopause* when periods have stopped for a year or more.

If you imagined that menopause is some grim announcement of the arrival of later life, think again. I would suggest that it is Mother Nature's way of protecting you. After all, hormones can be dangerous. As we have seen, too much estrogen exposure increases your risk of cancer and other serious problems. So Nature gave us a time-limited window of fertility so as to protect our overall health.

Some have suggested that menopausal symptoms are the artificial result of women living longer. Author Gail Sheehy wrote that in the early 1900s, a woman could expect to live only to the age of forty-seven or forty-eight.[1] So a woman living to fifty or fifty-five was already past her sell-by date. Problems are to be expected.

To this, I say hooey. There is simply nothing to it at all. First of all, average life expectancy was shorter in 1900 largely due to infant mortality from infections and other problems. Think of it this way: If, for example, a woman had two children, one of whom died in infancy and the other died at age eighty, the *average* life span—that is, the average of the two—would be about forty. Similarly, if half the population died of infections in infancy, and the other half lived to be eighty-four, the average life span would be about forty-two years—that's the average of people who die very young and those who live a long life. It does not mean that you are likely to die at forty, forty-two, fifty, or anywhere around then. Indeed, a century ago, mortality in infancy and early life was far greater than it is today. But if you got through the rigors of infancy unscathed, your chances of living to old age were actually very good.

Pregnancy-related mortality was also greater a century ago than it is today, and there were more fatal infections at any stage of life. But the notion that our bodies are not designed to last more than four or five decades is simply wrong. There have always been women and men living to a ripe old age. Benjamin Franklin lived to be eighty-four. Women's rights activist Susan B. Anthony lived to be eighty-six. Abolitionist Harriet Tubman lived to be ninety-one. None of them had the advantages of modern drugs or the latest surgical procedures. Their life spans were normal.

So menopause somewhere around age fifty has nothing to do with old age or the end of life. It simply means that you are now at an age when you no longer have a toddler in your household. Life has lots more in store for you.

But what about hot flushes and all those other symptoms? What causes them? Fascinating answers came from a look at other countries. In the 1980s, medical anthropologist Margaret Lock of McGill University interviewed thousands of women between the ages of forty-five and fifty-five, living in Canada and the United States, and compared them with women in Japan. The Japanese part of her study included 1,225 women living in Nagano, Kyoto, and Kobe.[2]

The differences were striking. Hot flushes and night sweats were rare in Japan. There was not even a Japanese word for them. Other menopausal symptoms—irritability, depression, and trouble sleeping—were similarly uncommon. For many women, menopause meant nothing at all, apart from the cessation of periods. The main symptom in Japan was shoulder stiffness, surprisingly enough, and men reported it about as often as women.

Is it that Japanese women were just too shy to complain about hot flushes? Dr. Lock investigated that possibility in interviews with more than thirty Japanese physicians and confirmed that Japanese women really were much less likely to have hot flushes and other menopausal symptoms, compared with their North American counterparts. These symptoms were not completely unknown, but they were uncommon. When they occurred, they were generally mild.

Why the differences? The most obvious explanation was food. As we saw in Chapter 3, "Tackling Cancer for Women," the traditional Japanese diet was nothing like the meat-based diet that has long been common throughout the United States, Canada, and Europe. The traditional Japanese diet was based on rice. There was relatively little meat and essentially no dairy products, until Westernization escorted these foods into Japanese culture. That meant that Japanese women were not exposed to dairy estrogens, and their foods had much less fat and more hormone-taming fiber than North Americans saw on their plates. This mostly plant-based diet also meant that Japanese women were slimmer, on average, than women in the United States and Canada. That meant less body fat to create hormones.

The Japanese diet also included plenty of soy-based foods: miso soup, tofu, tempeh, and others. What this means is that the traditional Japanese diet kept hormones in better balance throughout women's reproductive years, making the hormonal changes at menopause much less dramatic. At least that is the best explanation we have.

Japan is not unique. A study of women in northern China in the 1990s found that hot flushes and night sweats were uncommon there, too, although many women did report other menopausal symptoms: irritability, backache, depression, and tiredness.[3] In other Asian countries, the menopausal transition has similarly been found to be much less problematic than in North America.[4]

In Mexico, decades ago, researchers interviewed 118 postmenopausal Mayan women, finding that hot flushes and cold sweats were simply not a part of the female experience.[5] One might wonder if perhaps early marriage and many pregnancies had adjusted their hormonal balance in some helpful way. Maybe so, but it also has to be noted that traditional Mayan dietary staples are simple, plant-based foods, especially corn and beans, rather than meat or dairy products. Like women in Japan on a traditional diet, women in Mexico's Yucatan Peninsula got lots of fiber, while fatty foods and dairy hormones were not the commonplace things they are today.

These observations suggest that when women follow a mostly or entirely plant-based diet during their reproductive years, the menopausal transition is more gentle.

Tragically, these nutritional advantages are gradually being erased. With dairy products entering the culture, meat-eating increasing, and rice consumption falling—Japanese waistlines are expanding, and diabetes and cancer have exploded. The experience of menopause is changing, too. A 2005 study found that hot flushes were reported twice as often as they had been in the 1980s, although they were still less common than in North America.[6]

Similarly, the Yucatan Peninsula of the mid-1900s has been transformed into a tourist destination. Beans and tortillas are still sold in restaurants and stores, but meaty, cheesy tastes have invaded along with planeloads of vacationers arriving at Cancún's busy airport.

When Menopause Became a Diagnosis

In North America, many people no longer think of menopause as a natural stage of life. It has become a diagnosis, due in large part to the pharmaceutical industry.

In 1941, *Premarin* was first marketed in Canada for the treatment of hot flushes. It was introduced in the United States the following year. Despite the fact that the drug comes from horse urine, American women were lured to it by promises of remaining young, vibrant, sexual, and wrinkle-free.

Premarin's sales pitch was articulated in *Forever Feminine*. Written by New York gynecologist Robert A. Wilson and published in 1966, the book described his attempts to "cure" menopause and give women "age-defying youthfulness." To Dr. Wilson, that meant "straight-backed posture, supple breast contours, taut, smooth skin on face and neck, firm muscle tone, and that particular vigor and grace typical of a healthy female," as well as vaginal health so that a woman could "fulfill all her wifely functions" more or less indefinitely. In his words, "menopause must at last be recognized as a major medical problem in modern society." The answer, he wrote, was to be found in hormone pills.

Dr. Wilson started with sheep ovary extracts but encountered too many allergic side effects. Stilbestrol, the synthetic estrogen, had side effects, too. Estradiol benzoate was better tolerated, but it was an injection and not very convenient. Finally, the arrival of Premarin changed everything. A convenient oral drug, it was "entirely free of side effects"—or so he thought. Hormone replacement therapy (HRT), using Premarin or similar drugs, gave women back the hormones that nature had foolishly taken away. With aggressive marketing, Premarin's popularity climbed, and in 1992, it became the number-one prescribed drug in the United States.

However, problems had already started to surface. In 1975, the *New England Journal of Medicine* published two articles linking HRT to cancer of the endometrium—the inner lining of the uterus. In one, HRT users had four and a half times the risk, compared with nonusers.[7] The other study broke things down by duration of use. Women who used HRT for just one to five years had a sixfold increased risk of endometrial cancer. For women using HRT for seven or more years, the risk increased fourteenfold.[8]

Sounds terrible, of course. But when prescribed to women who had had a hysterectomy—so they couldn't possibly get endometrial cancer—the drug looked more or less safe. The same was true when Premarin was prescribed along with a progestin—a progesterone-like drug designed to counter the uterine cancer risk. In observational studies, women taking Premarin and similar hormone preparations actually seemed healthier than other women. The drug was not just reducing hot flushes; it also seemed to protect against heart attacks and other health problems.

So, in 1991, the National Institutes of Health—the U.S. government's research branch—launched a huge research study. The Women's Health Initiative included more than 160,000 women. It examined many things—whether HRT could reduce the risk of heart disease and fractures; whether a low-fat diet could prevent cancer, heart disease, and stroke; and whether calcium and vitamin D could prevent hip fractures. In the hormone part of the study, a group of women took Premarin or Prempro (a mixture of Premarin and medroxyprogesterone acetate, an artificial progesterone), while other participants got a placebo—dummy pills that looked like Premarin or Prempro but had no hormones in them.

Before long, however, serious problems emerged. Women taking Premarin had strokes, dangerous blood clots, and dementia more often than the women taking the placebo. Women taking Prempro had all those risks, too, plus a higher risk of breast cancer and heart attacks. The researchers took a long, hard look at the mounting list of health dangers linked to Premarin and Prempro, and in 2002, they made the only decision they could make: They pulled the plug. They stopped the study and contacted the research participants, telling them to stop taking the hormone pills.

Overall, women taking the hormone combination ended up with a 29 percent higher risk of coronary heart disease, a 26 percent higher risk of invasive breast cancer, a 41 percent higher risk of stroke, and a doubling of their risk of blood clots in the lungs.[9]

A look at Premarin's prescribing information today is enough to make any doctor drop his or her prescribing pen. Using Premarin alone, there is an increased risk of endometrial cancer (except for women who have had a hysterectomy), stroke, dangerous blood clots, and dementia. Using Prempro, there is increased risk of stroke, heart attacks, breast cancer, blood clots (including potentially fatal blood clots in the lungs, called *pulmonary embolism*), and dementia.

After the Women's Health Initiative findings became known, HRT sales dropped 63 percent.[10] And in the study's wake, something remarkable happened. As women said no to Premarin, breast cancer rates started to fall, too. A 2014 analysis showed that, because of the dismal results of the Women's Health Initiative, 4.3 million fewer women were using HRT. In turn, the drop in HRT prevented an estimated 126,000 breast cancer cases and 76,000 cases of cardiovascular disease.[11] Women were no longer sub-jecting themselves to the dangers of these commercial hormone products.

How did doctors react? You might think they would have embraced the findings and might stand up and cheer for the drop in breast cancer that came from so many women throwing Prema-rin in the trash can. But that is not what happened. An American College of Obstetrics and Gynecology survey showed that most member physicians were skeptical of the Women's Health Initia-tive results and continued to believe that Premarin was generally safe. That was especially true for male physicians, those who had been out of training for a longer time, and those practicing in Southern states. Women physicians, those who had recently com-pleted their training, and those practicing on the East Coast were more likely to accept evidence of HRT's dangers.[12]

Needless to say, patients were nervous about HRT. Many requested alternative treatments. And many felt they could no lon-ger trust their physicians' advice about HRT.[13]

In 2010, the Women's Health Initiative was able to report on eleven more years of follow-up. Indeed, women taking Prempro were more likely to develop breast cancer and to die of it, com-pared with women taking a placebo.[14] By 2017, the risks had evened out. Having stopped the hormone treatments many years earlier, deaths from heart disease or cancer were about the same in both groups.[15] But overall, HRT's risks remain. On December 12, 2017, the U.S. Preventive Services Task Force weighed in, recommend-ing against the use of HRT for the prevention of chronic health problems.[16]

Premarin's manufacturer is now fighting for a comeback, encouraging women to take it, at least for a little while, and is trying to find evidence of some health advantage or another. The drug is still advertised and heavily promoted, and it still finds plenty of cheerleaders in the medical community.

"Premarin" Is Short for Pregnant Mare's Urine

Apart from Premarin's health risks, there is another reason to think twice before filling a prescription. Behind the wildly exaggerated promises of youthfulness and femininity are some ugly facts about its source. The name "Premarin" is a shortening of the words "pregnant mare's urine": Pre-mar-in. The drug company uses the urine of horses who are impregnated annually, causing them to produce large amounts of estrogen.

The horses themselves do not have an enviable time. The mares are impregnated every June or July, either by stallions kept for that purpose or by artificial insemination. Then, in the fall and early winter, they are confined to small stalls so that their urine can be gathered with a special collection device. The stalls are narrow—just three and a half feet wide for smaller horses and five feet wide for larger horses.

Pregnant mares eventually give birth, of course, and if you imagined a bucolic scene of mothers and babies galloping through the hillsides, the fact is that their foals end up in the urine-production trade, are fattened up for the horse meat trade, or are sold to various ranching and sporting facilities.[17]

Better Choices

A woman who is turned off by Premarin will find plenty of other choices. First of all, other hormone brands are *not* made from horse urine. Estrace, Estraderm, Cenestin, Ogen, Ortho-Est, and many others are derived from plant sources (e.g., yams or soy) and are sculpted into hormones that are a match for your own.

Some women take "bioidentical hormones," hoping for something safer than typical prescription drugs. "Bioidentical" means hormones that have the same chemical structure as your natural hormones. In other words, the main estrogen in a young woman's bloodstream is 17β-estradiol, and that is what a "bioidentical" prescription promises you. Similarly, bioidentical progesterone is presumably just progesterone—the same as the hormone your body makes. In contrast, Premarin has some 17β-estradiol, but it also has equilin and 17β-dihydroequilin, which are horse estrogens, not human estrogens. And Prempro mixes horse estrogens with a progesterone-like compound that it is not identical to the progesterone your body makes. It has been modified so as to be absorbable when taken orally.[18]

Bioidentical hormones sound like boutique products, and, indeed, some are made-to-order mixtures compounded by special pharmacies that are prescribed in various regimens as selected by your physician. But the fact is, many common FDA-approved prescription hormones sold at ordinary drugstores are exact copies of human hormones, too. Estrace, for example, is a plant-derived copy of human estradiol. Ditto for Climara, Estraderm, and many others.

Because they do not involve the unsavory aspects of the horse-urine trade, they are certainly more appealing than Premarin. It is not yet clear that they are safer. One might hope that the problems of Premarin and Prempro could be blamed on the fact that they contain horse compounds that are not matches for human hormones, and that hormones that match the ones your body makes would be safer. But keep in mind that your own natural hormones can work mischief, too. As you'll recall from Chapter 3, researchers found that postmenopausal women with higher levels of estradiol in their bloodstreams had double the risk of breast cancer, compared with women with lower levels.[19] In other words, hormones have risks, no matter where they come from. Continued exposure to hormones—even hormones that are identical to your own—year after year is likely to present risks.

This is not just true of estrogens. Your body makes *growth hormone* to help you grow, as its name suggests. But people whose bodies make extra growth hormone end up with a condition called *acromegaly*, with abnormal growth and many health problems, including a higher-than-normal risk of colorectal cancer. Similarly, your body makes thyroid hormones, but having too much of them in your system can be deadly. The point is, just because something is a match for what your body makes does not mean that more is better.

While advocates of bioidentical hormones often describe menopause as a colossal goof of nature that medical science needs to rectify with pharmaceuticals, their safety remains in doubt. I hope that evidence will show these hormone products to be safe, but so far that has not been the case.

Other Approaches

There are other medications for menopausal symptoms, including selective serotonin reuptake inhibitors (e.g., Brisdelle), clonidine (a blood pressure medication that also reduces hot flushes), and others. And there are answers to menopausal symptoms that have nothing to do with drugs. Let's take a look at common problems and how to address them.

Hot Flushes

You break into a sweat, fan yourself, and in a few minutes, it's over. Then you get chills. Not all women experience hot flushes, but many do. And while doctors sometimes prescribe HRT for hot flushes, there are good reasons to say no to drug treatments.

First of all, hot flushes will typically go away on their own. It will take time, but for most women they will eventually stop. Second, for many woman, HRT's relief may be only temporary. When the hormone treatment stops—as it must for safety reasons—hot flushes sometimes kick in anew, and all you have done is delay them.

Third, as we have seen in detail, HRT is risky, something we learned the hard way. It increases your risk of heart attack, stroke, breast cancer, blood clots, and dementia. Here are safer ways to tackle hot flushes:

Trim away extra weight. You might imagine that having a little extra weight ought to protect against hot flushes. After all, body fat produces estrogens, as we saw in Chapter 1, "Foods for Fertility," and that extra estrogen ought to help against hot flushes. However, research has shown exactly the opposite. In the Study of Women's Health Across the Nation (SWAN), which followed 3,302 women over a ten-year period, thinner women were less likely to be bothered by hot flushes.[20] Is that just because it's easier to keep cool if you do not have an extra layer of body fat? Or is it some unusual effect of hormones? We don't know, but it really does pay to trim excess weight. In Chapter 12, "A Healthy Diet," I'll show you the best ways to go about it.

Favor soy products. Speaking of which, soy isoflavones have been shown to reduce hot flushes. As we saw earlier, miso soup, edamame, tofu, tempeh, and other soy products are popular in Japan. So a 1999 study in Gifu in central Japan compared the diets of women who had hot flushes with those of women who did not. The women who had avoided hot flushes tended to eat more soy, especially fermented soy products, such as tempeh.[21] Next, the same researchers followed a large group of Japanese women year by year as they went through menopause.[22] It turned out that the women who included the most soy products in their routines were 68 percent less likely to have hot flushes, compared with those having the least soy products. The idea is that having soy products in your routine before menopause will help prevent them from happening.

But what if you are having hot flushes now? Will soy products help? Australian researchers answered that question in a twelve-week study in women who were having hot flushes at least twice a day. By using a special soy flour that was added to drinks, mixed into cereal, or baked into muffins, hot flushes became less frequent within the first few weeks and, by the twelve-week mark, had dropped by 40 percent, very much like Ann's experience, as we saw earlier.[23] Various studies have had mixed results, and not every study has shown a benefit.[24,25] Overall, however, soy products do seem to help, but do not entirely eliminate hot flushes.[26]

Various herbal treatments are promoted for hot flushes, too. Overall, their effects are modest.[27] Red clover has shown some benefit, particularly for women with frequent hot flushes, using an 80 mg dose.[28,29] Black cohosh has also been tested, with mixed results. A recent study using 20 mg doses showed a reduction of hot flush frequency and severity, with the improvement gradually kicking in over an eight-week period.[30] You can find them online and in health food stores.

Sexual Functioning

After menopause, the normal drop in estrogen means the vaginal lining becomes thinner and the vaginal canal becomes shorter and narrower, often leading to changes you can feel: dryness, itching, and pain during intercourse. Like the other natural phases of menopause, this, too, has been given medical labels: *atrophic vaginitis, vulvovaginal atrophy,* or the current name, *genitourinary syndrome of menopause.*

For sexually active women, ospemifene, sold under the brand name *Osphena,* is an oral *selective estrogen receptor modulator,* meaning it acts like an estrogen for vaginal dryness and painful intercourse. Its risks are presumed to be similar to oral estrogens. Lubricants and moisturizers sometimes help. Local estrogen preparations are typically more effective, because they restore the vaginal tissues. There are many available products, such as Vagifem vaginal insert (small dissolving pills), Estrace cream, Estring vaginal ring, and others with plant-derived estrogens. Imvexxy is an estradiol vaginal soft-gel insert that comes in high- and low-dose formulations. Premarin's manufacturer also sells a vaginal cream for this purpose, but it is derived from horse urine, with the negatives that we saw earlier. Intrarosa is a vaginal insert of prasterone, a synthetic compound that converts to androgens and estrogens inside the vagina, although its mechanism of action is not fully established.

A few caveats: First, safety remains an open question. Vaginal estrogen products release traces of estrogens into the bloodstream, which may be of particular concern for women who have been previously treated for cancer. However, because they are used locally, their systemic effects are almost certainly less than those of pills or patches. Additional reassurance comes from a 2018 report from the Women's Health Initiative. In this case, the researchers were not looking at the use of Premarin or Prempro, which had shown serious risks. They were now looking at women who used vaginal estrogens—whatever brands they might have chosen. And it turned out that, over a seven-year follow-up period, women using vaginal estrogens had no increased risk of cancer or heart disease.[31] So far, so good.

Second, although these drugs make it *possible* for you to have sex, they won't necessarily make you *want* to. In other words, they may not boost your libido. The fact is, it is natural that the preoccupation with sex that is characteristic of adolescence and young adulthood will loosen its grip on your psyche as the years go by. While Dr. Wilson might have felt that a woman has a duty to "fulfill all her wifely functions," that's really up to you.

Inventing Diseases to Sell Drugs

Adriane Fugh-Berman, MD, of Georgetown University Medical Center, points out that the pharmaceutical industry has managed to turn natural changes in libido into a disease demanding a drug treatment. In 2004, industry came up with the diagnosis "hypoactive sexual desire disorder" as a way of marketing a testosterone patch for women.[32] Clinicians were to look for "persistently or recurrently deficient (or absent) sexual fantasies and desire for sexual activity" that causes "marked distress or interpersonal difficulty." As Dr. Fugh-Berman points out, "you might be perfectly happy with your libido, but if your partner isn't, you can still be diagnosed with a disease."

Men are targeted, too. As soon as testosterone was commercially available in gel form, the pharmaceutical industry began airing television commercials pushing men to be tested for "low T." While very few men actually need testosterone treatments, the number of men who can be induced to accept a prescription once they have been given a diagnosis is huge, as is the number of doctors willing to uncap their prescribing pens. Websites encourage men to suspect low T if they have symptoms as vague as sadness, grumpiness, or lack of energy—conditions that are more likely to be due to overwork or lack of sleep than hormonal changes.

Third, with vaginal creams and tablets, a small amount of the hormones can be absorbed by your male partner during sex, although this has not been shown to have any adverse effect.[33] It appears that absorption can be minimized by avoiding sex immediately after inserting a cream or tablet.

You may wish to explore the new nonpharmaceutical treatments now available through gynecologists. MonaLisa Touch delivers fractional CO_2 laser energy to soft tissues in a simple office procedure that does not require anesthesia. ThermiVa uses radiofrequency energy to rebuild collagen that may have been lost in the aging process. The vFit uses low-level light, gentle heat, and sonic technology to encourage blood flow through sensitive tissues. All are simple and safe.

Soy isoflavones have been shown to relieve vaginal dryness for some women.[34] Many other products have been investigated, including pueraria mirifica gel, fenugreek husk (*Trigonella foenum-graecum*), maritime pine bark (pycnogenol), red clover, and fennel. Studies have generally been few and far between with small numbers of participants, making it challenging to assess their efficacy.[35,36] Nonetheless, you will find these products at health food stores and online.

Mood and Cognition

As many women have discovered, the hormonal shifts of menopause can cause moodiness, irritability, and even depression. Depression strikes women more frequently than men anyway, and it is especially likely to arrive at menopause. By "depression," I do not mean feeling down in the dumps for an afternoon and then bouncing back to your previous self. I mean a continuing period of depressed mood, accompanied by poor sleep and poor appetite (or sometimes the opposite—*increased* sleep and *increased* appetite), as if your brain has shifted into low gear.

A University of Pennsylvania study followed a group of women over an eight-year period as menopause approached, asking them to fill out questionnaires about how they felt. Symptoms of depression were four times more common and a diagnosis of full-blown depression was more than twice as likely during menopause, compared with the years preceding it.[37]

The fact is, hormone shifts influence the mood-regulating neurotransmitters in the brain. Along with mood issues can come memory problems. They are usually minor and transitory, but can be annoying. In Chapter 11, "Foods That Fight Moodiness and Stress," we will tackle mood issues in detail, focusing especially on an exciting scientific frontier—the role of foods in controlling how we feel from day to day.

Part II

HORMONES, METABOLISM, AND MOOD

CHAPTER 7

Curing Erectile Dysfunction and Saving Your Life

Our research participants gathered for their weekly meeting. It was 6:00 p.m., and everyone had arrived from work or wherever they had been during the day. As usual, they stood on the scale to track their weight loss and then settled into the group room for a discussion of the past week's successes and challenges.

The research study focused on diabetes. The participants had made big changes in their eating habits and were reaping the rewards.

Ray, one of our participants, spoke up: "I'm doing so much better. My medication doses are way down, and my nerve pains are more or less gone. And one other thing. Um, how do I put this delicately?" he said. He had everyone's attention.

"Well, last night it was like I was young again. My ED seems to be gone, too." High fives all around.

We have seen this sort of turnaround in sexual functioning in many men. It is not something you will hear in a television commercial, and your doctor has probably never mentioned it. But changing the way you eat may well cure erectile dysfunction (ED). Let me show you how it works.

Thinking Beyond That Little Blue Pill

Researchers at Pfizer were trying to produce a treatment for chest pain and hypertension. Despite some initial promise, Viagra turned out not to be so hot for those problems. But the research participants did notice one surprising side effect—and Pfizer cashed in big-time. Approved for use for erectile dysfunction in 1998, the drug has earned the company tens of billions of dollars.

However, ED is not caused by a "Viagra deficiency." It is often a sign of impaired blood flow, and a man who leaves his doctor's office with nothing more than a recommendation to take Viagra is headed for trouble. Some cases result from medications, such as antidepressants in the selective serotonin reuptake inhibitor category (e.g., fluoxetine [Prozac], sertraline [Zoloft], or citalopram [Celexa]). Prostate surgery can do it, too. But most cases are caused by artery disease that a man did not know he had and that Viagra does not treat.

The male sexual anatomy is a hydraulic system that relies on good blood flow in order to work properly. If your arteries are wide open and healthy, things work fine. If the arteries have been narrowed by atherosclerosis, you will not have the blood flow you need. You can take Viagra, but all it does is widen your arteries temporarily. It does nothing to fix damaged arteries.

This is a wake-up call. You very likely have the same problem in the arteries to your heart, your brain, and all the rest of you. You might be worried about ED, but a much more serious problem is staring you in the face. A middle-aged man with ED is at serious risk of a heart attack or stroke. Researchers have found that after the beginning of ED, symptoms of cardiovascular disease can be expected within two to three years.[1]

This finding helped doctors make sense of a curious fact: The risk factors for heart disease—high cholesterol, high blood pressure, diabetes, obesity, and smoking—are identical to the risk factors for ED. That is because they are two symptoms of the same underlying condition—damaged arteries.

You are perhaps familiar with atherosclerosis, but let me walk you through what it means and how it relates to ED. As cholesterol particles circulate in the bloodstream, some of them can irritate the artery wall, causing a bump to form in the inner lining of the artery. The bump is called a *plaque*. Atherosclerosis means that plaques are gradually forming inside your arteries. As they narrow the passageway for blood, they slowly choke off the blood supply to your organs. In your heart, this loss of blood supply is experienced as *angina*, or chest pain. Sometimes, a plaque can burst like a pimple, triggering the formation of a blood clot, which suddenly blocks the artery completely. This causes a heart attack—which doctors call a *myocardial infarction*. It is the death of a portion of your heart muscle. This can also occur in the arteries to your brain, causing a stroke. And when atherosclerosis narrows the arteries to your private parts, the reduced blood flow leads to ED.

Back Pain, Too

Perhaps surprisingly, the first place where atherosclerosis occurs for most people is actually not the heart. It is the lower back. As your aorta leaves your heart and passes downward in front of your spine, it sends arteries into each of your vertebrae and eventually splits to run down your right and left legs. And that's where atherosclerosis typically starts—in the aorta, just as it passes through the lower back. By age twenty, many people in Western countries have atherosclerosis in their abdominal aorta to the point that they have completely lost one of the arteries to the vertebrae. It has been paved over by this disease process.

In turn, that loss of blood flow means that the disks—the leathery cushions between the vertebrae—do not get the oxygen and nutrients they need and can become fragile. Eventually, the disks can break open. Like stuffing coming out of a pillow, the soft interior of the disk herniates outward, pressing on a nerve, causing severe back pain that can radiate down the leg.

The point here is that plaques grow in the arteries and one of their favorite places is in your lower back. Just below that, the aorta divides into the common iliac arteries that run down each leg. The common iliac artery leads to the internal iliac artery, which leads to the internal pudendal artery, which leads to the penile artery. Along the way, the arteries are, of course, getting smaller and smaller.

Your back health and your sexual function depend on blood getting through. If your diet and lifestyle cause atherosclerosis, which narrows these small arteries even further, you are now a candidate for back pain and the target audience for Viagra.

This does not have to happen. And if arteries have narrowed, this process can often be reversed.

Low T? Not Likely

Some men imagine that flagging sexual potency is a result of low testosterone, and there are plenty of television commercials that would love for you to take a prescription to the pharmacy counter. If you are feeling less energy, deriving less pleasure from sex, are having more trouble with performance, maybe your problem is "low T," they suggest. Testosterone will make you feel like a man again. That's the promise.

The ads certainly work, although the treatment may not. A 2017 *JAMA* article found that television ad campaigns had driven major increases in the number of men seeking testing for "low T," and many more beginning treatment without bothering to test at all.[2] Between 2000 and 2011, testosterone use in the United States increased nearly fourfold.

If your blood tests show you are actually low in testosterone, testosterone treatment typically has only a modest effect on libido, erectile function, and sexual satisfaction, and no effect at all on energy or mood.[3] If your blood testosterone test results are in the normal range, it will likely do nothing at all—at least, nothing good.

Will it hurt you? Hopefully not. A 2018 study linked testosterone treatments to a higher risk of heart attacks or strokes, but only in older men and only in those receiving injections, as opposed to other forms of testosterone.[4] The effect on prostate cancer risk is uncertain. Many men have small clumps of cancer cells in their prostates. It is not yet known whether treatment with testosterone will cause these largely dormant cells to become more aggressive.

All in all, "low T" appears to be the modern-day equivalent of "iron-poor blood"—the 1950s-era notion that low iron was the cause of fatigue and *Geritol* was the answer. There are, in fact, lots of reasons for fatigue and low sexual desire, and low iron and low testosterone are far from the most likely causes.

If you are considering being tested for "low T," be sure your doctor does not have "low J." That is, you need a doctor with good judgment who will not push pharmaceuticals when they are not indicated.

How to Open Narrowed Arteries

In 1990, Dr. Dean Ornish made medical history. In a carefully conducted research study, he aimed to see if heart disease can be reversed—that is, could arteries that have been narrowed by advancing atherosclerosis reopen again?[5] He used a program that went further than the usual "heart-healthy" diets based on chicken and fish that had never proven particularly effective; he used a combined approach with a vegetarian diet, regular exercise, stress management, and no smoking.

After a year, the research participants had an angiogram—a special X-ray that shows the trickle of blood that gets through to the heart—and the results were compared to the same test done when the study began. It turned out that the arteries were reopening so much that a measurable difference was seen in 82 percent of patients—with no medication and no surgery, just a simple program of lifestyle changes.

Here is the take-home message: If a healthy diet and lifestyle can reopen arteries in the heart, the same thing can happen in arteries throughout the body, which is why Ray's ED got better.

To be clear, here are the steps for reversing artery disease:

1. **A plant-based diet.** Plants have effectively *no cholesterol at all*. And obviously, they have no animal fat. That is key, because animal fat is high in *saturated fat*, which boosts cholesterol levels. Getting animal products off your plate gives your arteries a chance to heal.

2. **Moderate exercise.** If you have heart disease currently, let your health care provider guide you as to how much exercise is safe. For most people, a thirty-minute brisk walk each day (that is thirty continuous minutes, or an hour three times per week) is a good place to start. If you are not able to do meaningful exercise, due to severe joint problems or other health problems, you will be glad to know that you can still do well, even without exercise.

3. **Stress management.** This can assume many forms. Dr. Ornish recommended yoga and meditation, and you'll find your own best method. See what allows you to healthfully tackle the stresses of day-to-day life.

4. **No smoking.** Smoking is rough on arteries, just as it is on your lungs and all the rest of you. Obviously, tobacco is out, too.

If this sounds challenging, remember: It beats surgery, chest pain, and chronic impotence. Soon you'll come to appreciate how great it feels to be healthy again.

Just Losing Weight Helps

Researchers in Naples, Italy, wanted to see what simple weight loss could do for ED.[6] One hundred obese men joined the study. On average, they were about forty-three years old and weighed around 225 pounds. Most had high cholesterol levels, and they all had ED.

Half the men were helped to lose weight. Over the next two years, they trimmed their portions and started regular exercise, mostly walking. As time went on, they lost weight—about thirty-three pounds on average—and their cholesterol levels fell slightly, too. In the course of the study, about one-third of the men regained their sexual function. A control group was given no specific advice and, not surprisingly, made no great progress on any of these measures.

The researchers then looked at who did well and who did not. The men whose sexual function improved the most were those who (1) lost the most weight, (2) exercised the most, and (3) had the biggest drops in *C-reactive protein* (CRP). CRP is a blood test that indicates inflammation. It is often elevated in overweight people. Inflammation damages the blood vessels and interferes with normal blood flow. A plant-based diet reduces CRP.[7]

All in all, the intervention was fairly modest. I would suggest setting bigger goals, using a plant-based diet to reopen the arteries so as to protect the blood flow to your heart and brain, and to your private parts. But the study showed that even with simple changes, things happen.

Exercise Helps

Yes, as we have seen, exercise helps reverse artery disease and ED. It was part of Dr. Ornish's program and part of the Naples study, too. It also seems to help prevent ED in the first place. The Massachusetts Male Aging Study tracked a large group of men in Boston, ranging in age from forty to seventy. Men who stayed physically active or who started a new program of physical activity cut their risk of ED in half, compared with sedentary men.[8]

Epilogue

Viagra researchers made another interesting discovery. They found that the drug is less effective when taken with a fatty meal. A greasy dinner can delay the drug's action by about an hour and can reduce its blood concentration by 29 percent.[9] So to make your Viagra work better, should you have it with a low-fat, vegan meal? Well, if all your meals are low-fat and vegan, you may never need Viagra at all.

CHAPTER 8

Conquering Diabetes

The world of diabetes has been revolutionized. With a new understanding of what leads to this disease and stronger nutritional treatments than ever, we are no longer simply managing it. We are directly tackling its cause.

Bob

Bob was fifty-five years old, living in Virginia. He was a U.S. Marine, serving in Beirut and Grenada with the 22nd Marine Amphibious Unit. After a knee injury ended his military career, he became a professional wrestler, fighting under the name "D.I. Bob Carter," taking his name from the tough drill instructor "Sergeant Carter" in the TV show *Gomer Pyle, U.S.M.C.*

He was muscular and slim. But after his wrestling career, a not-especially-healthy diet began to take its toll, and he gained weight, ending up at 296 pounds. At age fifty-three, the inevitable happened. His doctor diagnosed type 2 diabetes. Bob knew exactly what this meant. He was at risk for a heart attack, vision loss, amputations, and kidney disease. At the VA hospital, he had seen patients lining up for dialysis, and he wanted no part of it.

"I had two sons, aged thirteen and fourteen, and a girlfriend I am quite fond of," he said. "That, coupled with the knowledge of how diabetics die, I swore I wasn't going out that way."

He called a friend to ask for advice. His friend said that the key was to change his diet and guided him to our research and that of others. He dug into the Internet to learn more, and he decided he was going to conquer this new opponent.

He had a long way to go. His hemoglobin A1C, the main test to track blood sugar control, was an unhealthy 9.9 percent (values of 6.5 percent or higher indicate diabetes).* His blood glucose was dangerously high at 288 mg/dl (normal levels are below 100), and his cholesterol was 207 mg/dl (normal values are below 200). None of this was good. His doctor started him on metformin to control his blood sugar. He did not tolerate its side effects, so he was switched to insulin injections.

And this was Bob's challenge. "Give me ninety days," he told his doctor. "I'm going to be off those shots." His doctor was skeptical, to say the least. Diabetes is a one-way street, and people just don't get off their diabetes medications—or so his doctor thought.

Bob studied our research findings on plant-based diets, which you will learn about, too. He devoured books and lectures and decided to jump in. Out with the bacon and eggs, and in with a healthy breakfast of oatmeal with berries and walnuts. Beans and greens figured big in his lunches and dinners.

Very soon, he noticed that things were changing. Pounds started coming off, and his energy was coming back. By the three-month

---

* Hemoglobin A1C (or just A1C for short) is a blood test that reflects your blood sugar control for the preceding three months. Values between 5.7 and 6.4 percent are considered "prediabetes." Values of 6.5 percent or higher indicate diabetes. For people with diabetes, doctors typically aim to keep A1C below 7 percent.

point, he had lost fifty-two pounds. His blood sugar had dropped from 288 to 86, and his A1C had dropped to 6.4 percent. As Bob had predicted, his doctor stopped his insulin. By the eight-month mark, he was down ninety pounds. On no insulin, his A1C had fallen to 5.9 percent. His total cholesterol was an amazing 125 mg/dl.

His most recent A1C was 5.3 percent—squarely in the normal range. "I feel like a million bucks," he said. "This lifestyle saved my life."

Today, he shares what he has learned with others. "I don't have an MD or a PhD," he said. "I have an LMT. I *love my toes*. I've got ten fingers and ten toes, and I aim to keep them."

Fourteen months after his doctors identified his diabetes, they took the diagnosis off his medical record. He did not have it anymore. It was as if his doctor stepped into the ring and held Bob's victorious hand up high. He had stared diabetes in the face and conquered it.

Guy

Guy is a real estate broker living on Long Island with his wife, Debra. Apart from carrying extra weight, he was in generally good health—or so he thought. But his vision had been deteriorating. He went to an eye doctor, which led to several other medical referrals. It eventually became clear that the visual problem was caused by diabetes, a condition he did not know he had. He was also referred to a cardiologist, who found more bad news. His heart had apparently been damaged several years earlier by a "silent" heart attack.

It looked like not-so-healthful foods had caught up with him. His grandfather, arriving from Sicily, had opened a bakery outside Brooklyn Heights, and the family's traditions included plenty of hero sandwiches with processed meats.

His doctor recommended a number of treatments, including metformin for his diabetes and stents for his heart. A stent is a small tube inserted into a coronary artery to keep it open. Because stents can lead to dangerous clotting, patients usually need blood thinners, and stents have other risks, too. Guy felt there had to be a better way.

Debra got on the Internet, looking first for ways to tackle diabetes. She learned about our approach and found it promising. From there, they found Dr. Caldwell Esselstyn, who wrote that heart disease could be stopped in its tracks by changes in eating habits. Specifically, he recommended throwing out animal products, oils, and flour products (e.g., bread and pasta). And that is what they did. They jumped into beans, rice, polenta, kale, and sweet potatoes. Over the next several months, Guy lost weight. He actually lost *a lot* of weight. He ended up giving away suits, sport coats, and pants that no longer fit. His A1C improved, too—so much that his doctor stopped his metformin.

When the dust settled, Guy's weight had dropped from 295 to 190. His A1C fell from 10.5 percent to 5.2 percent, on no diabetes medications. And with no medications for cholesterol, his latest total cholesterol was 128 mg/dl, with an LDL of 69 mg/dl. All of these numbers are great. Guy's vision—which is what started his medical journey—returned to normal. At the same time, Debra has lost twenty-five pounds and feels great, too.

Their kitchen is stocked with beans, rice, lentils, and greens, and Guy is thrilled. "There has been such a change," Guy said. "I will never, ever go back." And he has passed his knowledge on to others. One of his clients shared concerns about his wife's heart problems, among many other health issues, necessitating care in a nursing home. His client decided to follow Guy's lead, and his wife is now doing much better and has been able to leave the nursing home.

A New Approach to Diabetes

In 2003, the National Institutes of Health funded our research team to test a new approach to type 2 diabetes. At the time, a typical "diabetes diet" meant portion control. People with diabetes were advised to cut calories for weight loss and to limit carbohydrates so their blood sugars would not rise too high. But that sort of diet did not fit what we knew about nutrition. Looking around the world, researchers found that countries where high-carbohydrate foods, like rice and sweet potatoes, were everyday staples actually had very little diabetes.

In Japan before 1980, diabetes was rare. In adults over age forty, the condition was found in just 1 to 5 percent of the Japanese population. A rice-based diet kept the country healthy. But Westernization escorted meat, dairy products, and other unhealthy foods into Japanese culture, and, by 1990, diabetes prevalence had risen to 11 to 12 percent.

In the United States, striking findings emerged from studies of Seventh-day Adventists. The Adventist religion places an emphasis on health, and many Adventists follow vegetarian or vegan diets. Many do not, however, and that has provided a natural experiment about the health effects of different diet patterns. In the Adventist Health Study-2, which included 60,903 participants, diabetes was found in only 2.9 percent of those following vegan diets, compared with 7.6 percent of meat-eaters. Ditto for weight problems—the average BMI among those on vegan diets was a healthy 23.5 kg/m^2, compared with 28.8 kg/m^2 among meat-eaters.[1]

Diets that omit animal products looked pretty healthy. So our team decided to test a low-fat vegan diet in people with diabetes. In a small pilot study, it worked very well. We then recruited a large group of volunteers. On average, they had had type 2 diabetes for more than eight years before joining the study. Half the participants began a conventional diet—cutting calories and limiting carbohydrates. The other half began a diet that eliminated animal products altogether and kept fatty foods to a minimum but had no limits on calories or carbohydrates.

After twenty-two weeks, the conventional group did well. Focusing on those who made no medication changes, their A1C fell by 0.4 percentage points. In the vegan group, A1C dropped by

1.2 percentage points—three times more than the conventional group.[2] Without limiting calories or carbohydrates, they did far better. They also lost weight and improved their blood sugar and cholesterol levels. They had new hope.

The study started a revolution in diabetes treatment. It showed that food changes were much more powerful than people had imagined. In this chapter, I will show you how to put these findings to use. But first, let's back up and understand the basics. What is diabetes, and what is the best way to tackle it?

Understanding Diabetes

In diabetes, there is too much sugar—that is, glucose—in the blood. Glucose is your body's natural fuel. Just as gasoline powers your car, glucose powers your muscles, brain, and all the rest of you. There is nothing bad about glucose. The problem in diabetes is that glucose is building up in the bloodstream, rather than going into the cells of your muscles or liver, where it belongs.

Normally, glucose is escorted into your cells by *insulin*, a hormone made in the pancreas. Like a key in a lock, insulin attaches to the surface of a muscle cell and signals the cell membrane to let glucose inside. It does the same for your liver cells. If you have diabetes, however, your insulin is not working right.

Diabetes Types

Type 1 diabetes occurs when the insulin-producing *beta cells* of the pancreas have been destroyed. This damage is caused by antibodies in the bloodstream. Because the pancreas no longer produces insulin, the condition is treated with insulin, administered by multiple daily injections or an external pump device.

Type 2 diabetes is much more common than type 1. In type 2, the pancreas is still making insulin, but the cells of the body resist insulin's action. It is as if the "key" has trouble turning the lock, so glucose cannot enter the cell as easily. This is called *insulin resistance*. To compensate, the pancreas makes extra insulin, but eventually it can no longer keep up, and your blood sugar rises.

Gestational diabetes occurs in pregnancy and is similar to type 2. Although it goes away when pregnancy is over, it is a sign that type 2 diabetes could be waiting around the corner, unless a woman changes her diet and lifestyle.

Checking Your Blood Sugar Control

Your health care provider can check your blood sugar control with various tests. Here is what the numbers mean:

Fasting blood glucose should be below 100 mg/dl (5.6 mmol/L). A value between 100 and 125 mg/dl (5.6 and 7 mmol/L) is considered prediabetes, and any value above 125 mg/dl (7 mmol/L) is considered diabetes.

Hemoglobin A1C, or just A1C for short, provides an index of your blood sugar control for the preceding three months or so. It is expressed as a percentage, and you want your value to be 5.6 percent or less. Values between 5.7 and 6.4 are considered prediabetes. A value of 6.5 confirmed with two separate tests makes a diagnosis of diabetes.

Glucose tolerance testing can also be done to get a more detailed view of your blood sugar control. You'll swallow a sugary syrup and your blood will be checked over the next two hours. The idea is to see if your blood sugar spikes very high and also to see if your body responds by making insulin, the hormone that escorts glucose from your bloodstream into your cells. Your doctor will interpret the results for you.

Let's see how foods affect all three types:

Foods and Type 2 Diabetes

As we have seen, plant-based diets are highly effective, both for weight loss and for improving diabetes.

You might be thinking, since a vegan diet probably includes a lot of carbohydrate-rich foods—spaghetti, rice, sweet potatoes, beans, fruit, and many more—it ought to raise blood sugar. So why does it do exactly the opposite? How does a plant-based diet improve blood glucose control so dramatically?

To understand this, let's take a trip to New Haven, Connecticut, where Yale University researchers are looking inside the human body.[3] Using a high-tech scanning method called *magnetic resonance spectroscopy*, the Yale researchers examined research volunteers, finding microscopic fat particles inside their muscle cells. These fat particles are called *intramyocellular lipid* (*intra* = "inside," *myo* = "muscle," and *lipid* = "fat"). Similar particles are also found in liver cells.

It turns out that as these fat particles build up within cells, they interfere with insulin's ability to work, as we discussed briefly in Chapter 5, "Reversing Polycystic Ovary Syndrome." In other words, fat inside cells causes insulin resistance.

As an analogy, imagine what would happen if some prankster were to jam chewing gum into your front door lock while you were away. When you arrive home and try to insert your key, you discover you cannot open the door. There is, of course, nothing wrong with your key. It is just that your lock is filled with gum. You need to clean out the lock.

In diabetes, the muscle and liver cells are filled, not with gum, but with fat. When that happens, the insulin "key" no longer works very well. And the solution is to clean that fat away.

Where do these fat particles come from? A breakfast of bacon and eggs trickles its load of fat into your bloodstream. At lunchtime, a bologna and cheese sandwich contributes more. Greasy potato chips, salad oils, or a cheese pizza for dinner add even more. These fats end up in your cells. In contrast, a vegan diet has no animal fat. If oils are minimized, too, the fat inside muscle and liver cells starts to dissipate.

Through a series of scientific studies, we have tested a plant-based diet for people with type 2 diabetes in many different settings, and also in people with long-standing diabetes who were experiencing the pain of *diabetic neuropathy*. Neuropathy can be unforgiving, causing continuing pain, especially in the feet, and greatly impairing quality of life. We found out that a low-fat vegan diet reduces painful symptoms and sometimes makes them go away altogether.[4] The same diet changes help women with gestational diabetes bring diabetes under control and can help prevent a recurrence.

Let me give you the specifics of how to implement this healthful eating plan. There are three steps: Avoid animal products, keep oils to a minimum, and favor low–Glycemic Index foods. There are no limits on calories or carbohydrates. Remember, we are not focusing on *how much* you eat. Rather, we are focusing on what you eat. Let's look at these steps in more detail:

1. Avoid animal products. By setting aside meat, poultry, fish, dairy products, and eggs, there will be no animal fat in your meals at all. In the bargain, this step also eliminates cholesterol, since cholesterol is essentially limited to animal products. Your plate will be filled with vegetables, fruits, whole grains, and legumes, and all the wonderful foods they turn into.

If you are wondering why fish is on the bad list, just like beef, it is because fish really is like beef. Yes, there is some omega-3 ("good") fat in fish, but most of the fat in fish is not omega-3. Here are some numbers:

A serving of roast beef has 3.4 grams of saturated ("bad") fat. Saturated fat raises your cholesterol. It is also implicated in Alzheimer's disease. A serving of Chinook salmon has 3.2 grams of saturated fat—not much different from beef. These products also pack cholesterol itself—83 mg for a serving of beef and 85 mg for fish. For comparison, the same-sized serving of black beans has just 0.1 gram of saturated fat and no cholesterol at all. Brown rice, broccoli, sweet potatoes, and most other plant-derived foods are similarly low. That's a huge benefit for your heart. And for your cells, where fat causes insulin resistance, the difference is like night and day.

Going back to the Adventist data, when it comes to weight and diabetes risk, fish-eaters are about halfway between daily meat-eaters and vegans. In other words, avoiding red meat and poultry is a good idea, and avoiding fish, too, is even better.

2. Keep oils to a minimum. Vegetable oils are more healthful than animal fats, of course. Generally speaking, they are much lower in saturated fat. But when it comes to calories, all fats are about the same. Both animal fats and vegetable oils have 9 calories per gram, in contrast to carbohydrates, which have only 4 calories per gram. So for weight control or to reduce the fat buildup inside your cells, you will want to skip animal products and limit vegetable oils (you cannot eliminate vegetable oils entirely, nor should you; plants have traces of vegetable oils that are essential for health).

So it is a good idea to minimize *all* added fats, including vegetable oils. The key is to use non-oil cooking methods. We will review them in Chapter 12, "A Healthy Diet."

A few plant-based foods (e.g., nuts, peanut butter, and avocados) are naturally high in fat. I would suggest avoiding these foods for now. Yes, the quality of their fats is much better than for animal fats, but our goal for now is to get extra weight off and to remove the fat from your muscle and liver cells.

If you were to add up your fat in the course of a day, it should add up to about 20 to 30 grams. That is much lower than is typical in North America, and it will help you reach your goals. In selecting foods, the fat content of typical vegetables, fruits, grains, and legumes is very low, and you do not need to worry about them. For packaged goods, such as a frozen dinner or sauces, check the labels. I recommend choosing those with a fat content of 3 grams or less per serving. If you eat a lot of packaged foods, you will find this a bit challenging at first, because many products exceed this limit. But you will soon find good choices.

3. Favor low–Glycemic Index foods. The Glycemic Index (GI) is a simple way of differentiating foods that raise blood sugar quickly from those that are gentler on your blood sugar. Simple substitutions will help you favor low-GI foods:

- Instead of table sugar, have fruit. Fruits have a sweet taste, but most have a surprisingly low GI and are easier on your blood sugar.
- Instead of wheat breads, favor rye and pumpernickel. Something about the rye grain causes it to have less effect on blood sugar.
- Instead of typical cold cereals, have bran cereal or old-fashioned oatmeal.
- Instead of typical white potatoes, favor sweet potatoes.
- Beans and pasta—even white pasta—are surprisingly gentle on your blood sugar.

A menu built from our healthy four food groups—vegetables, fruits, whole grains, and legumes—is naturally low-GI.

You may have noticed that the diet changes recommended for diabetes are actually identical to those for PCOS, described in Chapter 5. That is because insulin resistance is central to both. By undoing insulin resistance, we can greatly improve these conditions.

Let me share a few additional tips:

First, let me encourage you to bring raw foods (e.g., fresh fruits and salads) into your routine. They tend to accelerate weight loss.

Second, be sure that you are not excluding healthy carbohydrate-rich foods, like sweet potatoes, beans, and fruit. It is true that limiting carbohydrate can reduce your blood sugar to a degree, because, as you know, carbohydrate releases glucose as it digests. However, the cause of insulin resistance is fat buildup, and tackling that is far more powerful for blood sugar control. If you are avoiding carbohydrate-rich foods, you may be increasing meat, dairy products, eggs, or oils, all of which can contribute to weight gain and insulin resistance.

Third, have fun. Browse through the wonderful recipes in this book, and see what is waiting for you. Enjoy new food products and new restaurant choices. As you experiment, you will have some great successes and the occasional dud. Don't worry; that is what experimenting is all about.

Foods and Type 1 Diabetes

The diet changes described earlier—avoiding animal products, minimizing oils, and favoring low-GI foods—are also helpful for people with type 1 diabetes. They will not eliminate the need for insulin, but they help in two important ways:

First, a healthy diet can reduce your insulin requirements substantially. Plant-based diets have not been rigorously tested for type 1 diabetes, but in individual cases, we have often found that insulin needs drop by 30 percent or more.

Second, a healthy diet protects your blood vessels. Keep in mind that the principal danger of diabetes is damage to the tiny blood vessels to the eyes, the heart, the kidney, and the extremities. So you want to baby your blood vessels. To do that, it makes sense to eliminate animal fat and cholesterol completely.

A healthy diet may also help *prevent* type 1 diabetes. As I noted above, type 1 diabetes is caused by antibodies that destroy the insulin-producing beta cells. In a classic study published in 1992, the *New England Journal of Medicine* reported an important breakthrough about where these antibodies may be coming from. Researchers had taken blood samples from 142 children with type 1 diabetes. Every single child had antibodies to proteins in cow's milk. Further study showed that these antibodies were capable of attacking the insulin-producing cells of the pancreas.[5]

Antibodies are supposed to target viruses and bacteria. As part of your immune defense system, that is their job. But antibodies can also arise in response to foods. Once they are launched, antibodies can do unintended damage. They can attack your joints, for example, causing rheumatoid arthritis. In the *New England Journal* study, it appeared that the antibodies sparked by cow's milk ended up destroying the insulin-producing cells in the children's pancreas, leaving them with type 1 diabetes. In other words, the children's pancreatic cells may have been the victims of friendly fire.

For now, this is still a theory, and more research is needed. However, this line of research opened up an important strategy for reducing the risk of type 2 diabetes. By breastfeeding infants and avoiding the use of animal milks, perhaps we can cut the chances this disease will occur.

Many people have grown up with the notion that cow's milk is a healthful beverage for children (or adults). I would encourage you to set this idea aside. Milk's biological purpose is to help a calf grow, so it is packed with saturated fat, sugar (lactose), and hormones, none of which your child needs. If you thought that nonfat milk was a great source of protein, it actually draws 60 percent of its calories from sugar (lactose), with more than 12 grams of sugar in a cup of nonfat milk. Children should be breastfed and the only beverage needed after weaning is water.

Avoiding Hypoglycemia

One important caution: A low-fat vegan diet is powerful. The combination of a low-fat vegan diet with insulin or medications that trigger insulin release can lead to *hypoglycemia*, or low blood sugar, which can be dangerous. This is the combination of a strong diet and strong medications.

This is not likely to occur if you are on no medications or are taking only metformin. But talk with your doctor about the medications you are taking and your hypoglycemia risk. Hypoglycemia can also occur following vigorous exercise in people taking insulin or sulfonylureas.

Here is how to protect yourself against hypoglycemia:

1. Let your health care provider know that you are making a diet change, and have a way to contact him or her in case of low blood sugar. In some cases, medications should be lowered before diet change, to prevent hypoglycemia. Work with your health care provider.

2. Know the symptoms of hypoglycemia:
   - Shaking
   - Sweating
   - Anxiety
   - Hunger
   - Weakness
   - Rapid heartbeat
   - Light-headedness or dizziness
   - Sleepiness or confusion
   - Difficulty speaking

3. Carry a glucose meter. If you experience the symptoms noted above, check your blood sugar. If it is less than 70 mg/dl (3.9 mmol/L), or whatever other value your health care provider has recommended, have something to raise your blood sugar quickly. If you are driving, pull over and check your blood sugar. If you cannot test yourself for whatever reason, assume you are low, and have something to eat. Then, check your blood glucose again after 15 minutes. If

your reading is still low, have another serving and check again in another 15 minutes. If you are near your usual mealtime, go ahead and eat. Otherwise, have a snack.

Here are good choices of foods to raise your blood sugar:

- Glucose tablets. You will find them at drugstores. Keep them in your desk drawer, your purse, and your glove compartment. If your blood sugar is indeed low, you should take 15 grams of glucose. For common glucose tablets, that means taking *four* tablets at once.
- ½ cup (4 ounces) of fruit juice. Any kind will do.
- ½ cup (4 ounces) of a soft drink. Do not choose a diet drink.
- 5 or 6 pieces of hard candy.
- 1 or 2 teaspoons of sugar.

4. Wear a medical identification bracelet.

5. Check your blood glucose during exercise.

6. Be attentive to the possibility of nighttime hypoglycemia. Yes, it can also occur while you are asleep. Here are common signs:
   - Heavy perspiration
   - Nightmares or crying out
   - Unusual tiredness, confusion, or irritability on awakening

If you suspect nighttime hypoglycemia, set your alarm for about 2:00 or 3:00 a.m. and check your blood sugar a few nights in a row. Share your results with your health care provider, who can adjust your medications as needed.

Hypoglycemia does not mean there is anything wrong with you or with your diet. It means that your diet is working, and that the combination of your healthy diet and your medications is now too strong for you. But hypoglycemia can be dangerous. It is important to let your caregiver know about your hypoglycemia right away. Otherwise, these episodes are likely to recur.

Lace Up Your Sneakers

Although good food is the centerpiece of a healthy lifestyle, exercise adds extra power. Exercise lowers blood sugar and can assist you in keeping weight off. Speak with your clinician about how much and what kind of exercise is right for you. If you get a green light, plan for two and a half to five hours of moderate-intensity physical activity (e.g., brisk walking, dancing) every week. If you have been sedentary, start slowly and work your way up gradually, and note the cautions above about checking your blood sugar.

If you can't do meaningful exercise, due to bad joints or another reason, a diet change will still help. The research studies that have shown the power of a plant-based diet were done *without* exercise.

Keep an Eye on Your Blood Sugar

A plant-based diet normally reduces blood sugar quickly and effectively. However, some people find that during the first few days of starting a vegan diet, their blood sugar values rise a bit. This is likely because you are still insulin-resistant, so starchy foods will cause your blood sugar to bump upward temporarily. Stick with the healthful guidelines above, and this will typically turn around in a few days.

If you have any kind of infection, such as a viral illness, it will raise your blood sugar, as will steroid medications. Your health care provider can help you sort out whether these might be the cause of blood sugar elevations.

If, for whatever reason, your blood sugar is just not budging, let's recheck the basics. Be sure your meals are completely free of animal products and added oils. Go on a search-and-destroy mission for fatty foods, such as nut butters and guacamole.

Avoid the temptation to limit your intake of healthful carbohydrates. Some people hope that by avoiding beans, sweet potatoes, and other starchy foods, their blood sugars will drop more quickly. Typically, the opposite occurs: People who try to avoid these foods have the *most* trouble over the long run. Keep healthful plant-based foods in your routine and let them do their work.

If you have not yet begun an exercise program, this is a good time to begin. Get your health care provider's okay before you embark on a new exercise program and work up to it gradually.

Occasionally, blood sugar control is resistant to lifestyle interventions. This suggests the need to be tested to see if the body is making enough insulin. In this situation, a healthful lifestyle is still very beneficial to prevent diabetes complications, but insulin may be required.

Sometimes the opposite can happen. Your blood sugar drops too low. If your blood sugars are dipping below about 70 mg/dl (or whatever other value your health care provider has recommended), check the guidelines above for hypoglycemia and contact your health care provider. It is likely time to reduce or discontinue your medications. A healthy diet alone does not cause hypoglycemia; this is the result of medications that are now too strong for you.

Feeling Like Yourself Again

If a diabetes diagnosis has kept you at the doctor's office and pharmacy counter, you are now freeing up your life to spend more time on the dance floor, the tennis court, a forest path, or wherever life takes you. Have fun.

CHAPTER 9

A Healthy Thyroid

The thyroid gland sits at the base of your neck, shaped vaguely like a butterfly. Despite its small size, it has a big job. It oversees your body's use of energy. That is to say, it maintains your body temperature and keeps your heart, brain, and muscles working right.

If you are a *Star Trek* fan, your thyroid is like Scotty, the engineer on the *Starship Enterprise.* Captain Kirk always called on Scotty for more power to get past one disaster or another. Sometimes Scotty could deliver. Other times Scotty had to respond with "Captain, we can't do it! If we keep this speed, we'll blow up any minute now!"

That's your thyroid. It gives you the energy you need—or at least it tries. It works by sending thyroid hormones into your bloodstream, reaching every organ in the body. If you are running low on thyroid hormones, you'll feel sluggish and cold. If your thyroid overproduces its hormones, you may feel warm and have a rapid pulse and other symptoms of being revved up.

Thyroid problems are very common. Let me share some real-life examples.

Nancy

Nancy is an accountant in Chico, California. In both finance and health, she keeps her eye on the numbers, and what happened to her numbers was remarkable.

At age nineteen, she began to suffer from unusual lethargy. At the time, she was studying at a junior college and working weekends at a veterinary clinic, and for no apparent reason, she just ran out of gas. As her energy flagged, it was all she could do to maintain her schedule. At the same time, her weight started creeping upward. Before she knew it, she was twenty-five pounds heavier than what was normal for her.

What was going on? At her mother's urging, she saw a physician to try to sort out the problem. The doctor suspected an underactive thyroid. Without adequate thyroid hormones, her metabolism would be slow, she would have no energy, and she would put on weight. He ran a number of blood tests and confirmed that she was indeed hypothyroid. He wrote out a prescription for levothyroxine—thyroid hormone—to make up for what her body was not producing on its own.

The medicine did not quite get her back to her old self. Although she was able to diet away some of those unwanted pounds, she did not return to her goal weight and never got her energy back. "I came to accept that being overweight and fatigued was just who I was destined to be."

Health problems were familiar to her already. They ran all up and down her family tree. "My father had a heart attack at forty-seven," she said. "He had bypass surgery twice and suffered from congestive heart failure. Eventually, he died of cancer." Her mother suffered from painful arthritis and developed breast cancer—as did three of her four sisters—as well as kidney cancer and Parkinson's disease. Nancy feared that she was following suit. "My future wasn't looking great. I was worried about my health in a very big way!"

In retrospect, she wondered if her health problems might have had their roots in the family's agricultural businesses. Her father's father immigrated from Oslo, Norway, and set up a dairy in California. Her mother's father raised Herefords. "As a child, I was active in 4-H and Future Farmers of America," she said. "We raised sheep and hogs and sold them at the fair."

Her parents had a walnut orchard. Surrounded by seventy acres of walnuts, they were routinely exposed to agricultural chemicals of all kinds. "And just like my parents, I ate the standard American diet," she said. Between not-so-healthful foods and risky chemical exposures, she could see the toll on her family.

In 2012, she happened to see a film called *Forks Over Knives*, which told the story of Colin Campbell and Caldwell Esselstyn, two researchers who had made remarkable discoveries of the power of plant-based diets to reverse heart disease, prevent cancer, and restore health.

"The moment the movie ended," she said, "I turned to my husband and said, 'I'm done. I'm going plant-based.' The next morning, I went into the kitchen with a big garbage bag and cleaned it out. I threw all the meat and dairy products away.

"I talked to my doctor about my plans and about my desire to get off all medications. But he said that was not possible. I would have to take medications for life. And so that became my challenge."

When her doctor retired, she started seeing a new physician who encouraged her to continue her plant-based diet and tracked her progress every six months.

"I was forty-nine years old and 265 pounds. So far, I've lost 105 pounds. My cholesterol went from 265 to 148 almost overnight—without taking a statin. My levothyroxine dosage went from 125 mg to 75 mg very quickly, and then we stopped the medication altogether. I have had more than two years of normal thyroid numbers. I'm no longer cold; my hands are warm. I have boundless energy, a thick head of hair, and healthy skin and nails. All the symptoms of thyroid disease are gone."

Her husband, Nick, joined her new approach to eating. He had weighed over 300 pounds and has lost 90 pounds. Both are off all medications and feel better than they have felt in years.

"We love greens and steel-cut oats with fresh fruit for breakfast. Lunch could be a big salad and soup. We keep it super simple and like to batch-cook. We have lots of cruciferous vegetables: broccoli, cauliflower, and Brussels sprouts, plus quinoa, potatoes, and wheat berries. And we love fruit. A plant-based lifestyle has been a true game-changer for my husband and me."

Mike

Mike is a healthy, active North Carolina neurosurgeon. In addition to his busy practice, he has long participated in competitive cycling, running, triathlons, tennis, and squash, and has been a picture of health.

Every year he has a routine physical, and most everything has always been fine. At age forty-five, however, blood tests showed a sluggish thyroid. Specifically, his TSH (thyroid-stimulating hormone) level was elevated. When the thyroid is not producing enough thyroid hormone, the brain releases extra TSH to try to turn on the underachieving thyroid gland. Values over 4 mU/L indicate hypothyroidism; physicians run additional tests to verify the diagnosis.*

The elevation was mild—his value was 4.2 mU/L—and maybe

---

* TSH is measured in milliunits per liter (mU/L). Normal levels are 0.4 to 4.0 mU/L. Higher levels suggest hypothyroidism; lower levels indicate hyperthyroidism. The diagnosis needs to be done by a physician, who will also check the level of thyroid hormone (T4) and run other tests.

it was a fluke. So he and his doctor decided to do nothing for the moment. But in the years that followed, his TSH levels inched upward to 4.7 and then to 5.1. With the same result cropping up after five years, he was clearly hypothyroid, and it was not going away. Perhaps it was time to fix it. That meant starting thyroid hormone replacement therapy.

Hmmm. Facts are facts, but Mike decided to wait. And as fate would have it, he had started to take an interest in nutrition at about the same time. While on a family vacation, he began reading about the advantages of getting away from animal products, and he decided to give a vegan diet a try. He started his day with oatmeal with flaxseeds, had bean dishes for lunch, favored fruits as snacks, made plenty of salads, and never had any animal products at all.

Soon, he noticed several things. Although his energy had not been too bad before, it was noticeably better. His skin looked healthier. He lost seven or eight pounds.

A few months later, he repeated his thyroid tests. And after years of consistent hypothyroidism, his TSH had dropped to 2.9, putting him back into the normal range. Not only did he feel better; his blood tests showed that indeed he *was* physically healthier.

Wendy

Wendy is a film producer in New York. At age forty-seven, she was working hard and was under more or less constant stress. Most nights she was up late, and she got up early every day. Often, she felt a bit "hyper," which she assumed was due to the demands of her work.

But she noticed other things, too. She gradually gained weight, ending up about thirty pounds heavier than she wanted to be, and she found it tough to lose. Her hair had become brittle and seemed to be falling out periodically. Worse, she had terrible PMS and mood swings, lapsing into serious negativity, especially around the time of her periods.

She needed to find out was wrong and went to her doctor for an exam. It turned out that her TSH level was very low, indicating that she was hyperthyroid. That is, her thyroid gland was working overtime. Often hyperthyroidism causes weight *loss*, but occasionally the opposite happens, as in Wendy's case.

Her doctor ran other tests and found high levels of thyroid antibodies—meaning that her immune system was attacking her thyroid. He sent her to an endocrinologist, who raised the possibility of giving her radioactive medications to destroy her thyroid before things progressed, saying that she would then need to take thyroid medications for the rest of her life. This was not a welcome prospect. Her doctor agreed to forgo any treatments for the moment and gave her three months to see if the situation might change.

Shortly after that, she went to a nutrition lecture, mainly to see what she could do for her mother, who had had an ongoing cholesterol problem. The lecturing physician promoted a plant-based diet, and she learned that it was not just good for lowering cholesterol; it could also help autoimmune conditions. She decided to give it a try.

Her diet had not been too bad, one might think. She ate no red meat, though she did have poultry, fish, and cheese. But over the next three days, she cleaned out her kitchen.

Never before had she received a prescription based on food, but that is exactly what she had now, and she followed it as strictly as she would have had the prescription been a medication. She went off all animal products, including dairy products and everything else. She went off gluten, too. She had vegetable and bean soups and stews, lots of greens, onions, mushrooms, seeds, and a salad every day.

After four weeks, she went back to her endocrinologist, who repeated the earlier blood tests. In the days that followed, she waited for a call with the lab results. But the phone never rang. So she called the doctor's office. The nurse told her, "We only call back if there is a problem." And Wendy said, "But I have had a problem." "No," the nurse told her. "Your results are normal." What? Could it be that her thyroid had actually returned to normal? They repeated the tests, and indeed she was totally healthy. Her thyroid tests were in the normal range. And by the six-week point, she had lost more than twenty pounds.

Shortly after that, her mother died, which threw her life off-kilter. She compromised her diet, and her thyroid antibody numbers worsened again. So she then got back on a whole-food, vegan diet and has been healthy ever since. Today, she steers clear of animal products, and also avoids gluten, flour, and sugar. She cooks delicious food with no oil and no salt. She has no thyroid problems and her moods are stable.

As we have seen, thyroid problems can lead to all kinds of health issues and can be challenging to sort out. It is important to have a good medical evaluation, and sometimes a diet change can make all the difference.

Let's look in more detail at what happens when the thyroid is misbehaving and how to give Scotty what he needs to do his job.

Hypothyroidism

An underactive thyroid can lead to fatigue, sensitivity to cold, dry skin, constipation, and weight gain. If things worsen, you might notice puffiness in your face, hoarseness, weakness, joint and muscle aches, hair loss, depressed mood, and noticeable gaps in your memory. Your cholesterol level can rise and menstrual periods can become heavy and irregular. Your thyroid can also enlarge—which doctors call a *goiter*—as your thyroid struggles to improve its hormone output.

In the most serious cases of hypothyroidism—which thankfully are rare—it is as if the body is turning itself off. Body temperature and blood pressure fall, breathing slows, and you can even lapse into a coma and die.

Why would anyone ever run low on thyroid hormone? World-wide, the most common reason is a lack of iodine. Iodine is an essential part of thyroid hormones; without it, the thyroid cannot do its job. In the United States and many other countries, iodine deficiencies are rare, thanks to iodized salt. In these countries, the most common cause of underactive thyroid is an autoimmune condition in which antibodies attack your thyroid gland.

Let me explain: When your body senses an invader—a virus, for example—your white blood cells make antibodies, which are like protein torpedoes that attack the interloper. Sometimes, your white blood cells mistakenly make antibodies *against your own body tissues*. For whatever reason, the thyroid is a common target. Antibodies that attack your thyroid cells interfere with their ability to function. Doctors call this *Hashimoto's thyroiditis*, after Hakaru Hashimoto, the Japanese physician who first described the condition in 1912.

Another reason for hypothyroidism is surgery. Some people have their thyroids removed because of thyroid cancer or are treated with radiation because of hyperthyroidism. Without any functional thyroid gland left, they end up hypothyroid and will need hormone supplements. More rarely, medications (e.g., lithium) and congenital conditions can also lead to hypothyroidism.

Hyperthyroidism

If your thyroid produces too much thyroid hormone, the effects are more or less the opposite of hypothyroidism. Instead of feeling cold, an overactive thyroid can make you feel too warm. Instead of gaining weight, you are losing weight without intending to (although in rare cases, weight gain can occur, as it did for Wendy). You may also experience a rapid or irregular heartbeat, a tremor in your hands, nervousness, irritability, weakness, and difficulty sleeping. Your hair can become fine and brittle, and your skin can become thinner. Women may find their periods are lighter or less frequent, and bowel movements can become more frequent. Although you may feel energized in the early stages of the disease, fatigue will be more prominent later on. In some cases, the body tissues behind your eyes can swell, pushing your eyeballs forward, making them protrude, although this is uncommon.

In the same way that antibodies cause hypothyroidism, they are also the main cause of hyperthyroidism. In this case, they are overdriving the thyroid, making it produce too much thyroid hormone, a condition called *Graves' disease*. It runs in families and, for some reason, is more common in women than men. A less common form of hyperthyroidism results from nodules within the thyroid overproducing thyroid hormones.

A doctor who suspects hyperthyroidism will feel for enlargement in your neck, look at your eyes, and check for a rapid or irregular pulse, a tremor, and overly quick reflexes. Blood tests will confirm the diagnosis. Doctors treat hyperthyroidism with anti-thyroid medications, radioactive iodine, or surgery.

What's in a Name?

The name *Graves' disease* does not mean the condition is especially dire. It can be, but the name actually comes from Robert Graves, an Irish doctor who described a case of goiter and protruding eyes in 1835. Graves was also an inventor, who originated the idea of a second hand on watches, presumably so he could take a patient's pulse.

In Europe, the condition is more commonly called Basedow's disease, after German physician Karl Adolph von Basedow, who described it in 1840. Several others have had their names tacked onto the disease, too: English physician Caleb Parry, Scottish physician James Begbie, Irish physician Henry Marsh, and Italian physician Giuseppe Flajani.

Understanding the Blood Tests

Your doctor can check your thyroid with simple blood tests. Here are the basics:

TSH, or thyroid-stimulating hormone, also called *thyrotropin*. If your thyroid is sluggish, the pituitary gland at the base of the brain will make extra TSH to try to get your thyroid moving. So high TSH levels indicate hypothyroidism, in most cases. If your thyroid is overactive, it is just the opposite; your pituitary doesn't want to push it to make more hormones, and your TSH level falls. So a low TSH indicates hyperthyroidism. The normal range for TSH is between 0.4 and 4.0 mU/L, although some have called for stricter limits.[1]

T4, or thyroxine. Think of T4 as the storage form of thyroid hormone that your thyroid gland releases into the blood, ready to turn into the more active form, T3. A high level typically means hyperthyroidism, while a low level indicates hypothyroidism. Certain conditions (e.g., pregnancy, oral contraceptives, or steroid medications) can alter your T4 test by changing the amount of proteins in the blood that bind T4. Doctors can also measure *free T4*—that is, T4 that is not bound to proteins and is free to enter the body tissues. The normal T4 range is 5.0 to 13.5 mcg per deciliter.

T3, or triiodothyronine. The normal range is 100 to 200 ng/dL. Sometimes in hyperthyroidism, the T4 level is normal, while T3 is high.

Laboratory tests can also check for the thyroid antibodies that cause hypo-or hyperthyroidism, and doctors have additional tests to check your thyroid health.

How Do Foods Affect the Thyroid?

A monarch butterfly lives on the nectar of flowers, and the butterfly in your neck has its own ideas about food, too. The thyroid's relationship with food is actually a bit different from that of other organs. Here is what you need to know:

Iodine

As we saw earlier, your thyroid uses iodine to make thyroid hormones. As you might have guessed, T4 has four iodine atoms, and T3 has three.

You need only a tiny amount of iodine—the recommended dietary allowance (RDA) for adults is just 150 mcg per day. But even getting that tiny amount can be challenging, since the traces in most foods are very modest. So many governments have turned to iodized salt as an easy solution. Morton Salt began offering iodized salt in 1924, leading not only to better iodine status, but also to a measurable increase in IQ in affected areas of the United States.[2]

One-third teaspoon of iodized salt gives you a day's worth of iodine, more or less. If you buy kosher salt, sea salt, fleur de sel from Brittany, or Himalayan salt from the Punjab, don't count on it having much iodine unless the label says it is iodized.

An especially healthful source of iodine is seaweed—or, to use a more respectful name, sea *plants*. They concentrate it from the ocean. A typical 2.5-gram sheet of dried nori (used for sushi) has about 40 mcg of iodine; a 1-gram serving of wakame (used in soups and salads) has 42 mcg. In Japan, where sea plants are everyday staples, typical iodine intake can reach 1,000 to 3,000 mcg per day, far more than anyone needs.[3]

Some iodine ends up in fish. It is also found in dairy products, due to nutritional supplements given to cows and the use of iodine-containing sanitizing agents on dairy farms. Traces are also found in common fruits and vegetables. Overall, however, it is easy to run low in iodine intake if you are not using iodized salt or including sea plants in your routine.

When it comes to iodine, however, more is not always better. Just as getting too little iodine can increase the risk for hypothyroidism, the same is true if you get too much.[4] So, it is a good idea to avoid iodine supplements, unless prescribed by your physician.

Immune-Friendly Foods

The most common cause of hypo- and hyperthyroidism has nothing to do with iodine. The most common cause—both for hypo- and hyperthyroidism—is an autoimmune attack. As we saw above, the immune system's torpedoes that are supposed to be hunting down viruses, bacteria, or other invaders inadvertently go after your own thyroid gland. The same thing happens in rheumatoid arthritis, when antibodies attack the lining of your joints, and in type 1 diabetes, in which antibodies destroy the insulin-producing cells of the pancreas. Just as antibodies are tough on viruses and other invaders, when they are misdirected, they can be tough on you, too.

So what triggers the release of antibodies? What is fooling your white blood cells into thinking that an invader is present and needs to be wiped out? Well, perhaps there really is an invader—a passing virus that aroused an antibody response that continues long after the virus is gone.[5] Immune responses can also be triggered by foods. If you think about it, a virus is tiny. But what you eat for lunch is huge—filled with foreign proteins that can get your immune system riled up.

To investigate the links between foods and antibody responses, researchers have compared people who had elevated levels of thyroid antibodies in their bloodstreams and those who did not. It turned out that those with higher antibody levels tended to eat more animal fats, particularly butter. Milk, eggs, processed meats (bacon, sausages, salami, etc.), tea, and oil were also prominent in their diets. And they tended to neglect vegetables, beans, fruits, nuts, and cereals.[6]

Could it be that a plant-based diet reduces the risk of thyroid problems, while an animal-based diet increases the risk? That certainly fits with the experiences of Nancy, Mike, and Wendy. Researchers in Loma Linda, California, tackled that question head-on. The Adventist Health Study-2 began in 2002. Taking advantage of the fact that some Seventh-day Adventists eat meat daily, while others follow vegetarian or vegan diets, the study set out to track the health of a large group of people following various dietary patterns.

The researchers had already found that food choices were linked to body weight, cholesterol levels, diabetes, and many other health issues. And now the researchers found that they were also related to thyroid disease. In 2013, having followed 65,981 participants, the research team reported that those who avoided animal products entirely (i.e., those following a vegan diet) were 22 percent less likely to become hypothyroid, compared with omnivores. In contrast, lacto-ovo vegetarians (people who avoid meat but eat dairy products and eggs) were *more* likely to become hypothyroid, compared with meat-eaters. Risk for the fish-eaters was in between. In other words, it looked like meat and dairy products were problem foods. People who avoided them did better.[7]

All of these findings could have been simply due to chance. However, they fit the pattern that we have seen for many conditions: Dairy products seem to trigger inflammatory conditions, while diets that exclude animal products in general have the opposite effect.

For hyperthyroidism, a similar pattern held, except that meat seemed to be a bigger problem than dairy products. Again, participants following a vegan diet had the least risk. They were 52 percent less likely to have hyperthyroidism, compared with meat-eaters. For lacto-ovo vegetarians, risk was reduced by 35 percent, compared with meat-eaters, and those avoiding meats other than fish had 26 percent less risk.[8] So, those avoiding all animal products did the best, meat-eaters did the worst, and the other groups were in the middle.

What About Soy Products?

Many people are choosing soy products these days. Soymilk and tofu used to be thought of strictly as Asian foods; today they are on grocery store shelves in Fargo, Omaha, Trenton, Corpus Christi, and everywhere else. People who cannot digest cow's milk do much better with soymilk, and some people choose soy products for their ability to lower cholesterol, prevent hot flushes, or reduce cancer risk.

Soy's effect on thyroid function has been explored in several studies, because some have raised the question as to whether soy could have an adverse effect on the thyroid. Overall, however, studies show that soy seems not to have much effect one way or another.[9,10,11,12,13,14] One observational study, the Adventist Health Study-2, showed no relationship between soy products and thyroid conditions in men, but did find higher TSH levels (meaning lower thyroid activity) in women consuming the most soy protein.[15] Like many other foods, soy products can also reduce absorption of thyroid medications. So you'll want to take your medications on an empty stomach.

All in all, these studies suggest that soy products do not adversely affect the thyroid. Even so, you will want to baby your thyroid gland with a healthy, plant-based diet and by being sure to include iodized salt or other iodine-containing foods in your diet on a regular basis.

If You Take Thyroid Hormones

If you are taking thyroid hormones (e.g., Synthroid), there is one more link between foods and the thyroid that is important to know about. Many foods and supplements reduce thyroid hormone absorption. These include coffee, calcium or iron supplements, soy products (as noted above), and just about any high-fiber food.[16,17] So, it pays to take your medicines at least thirty minutes before taking any food. Absorption is better on an empty stomach.

A Menu for a Healthy Thyroid

As we have seen, foods affect the thyroid gland in many ways. For some people, a diet change seems to have been a cure for thyroid problems. I have to say, this surprised me. This is clearly a new area and one that is ripe for solid research studies before we can draw firm conclusions. In the meantime, here are some simple tips for keeping a healthy thyroid:

Iodine: not too much, not too little. Iodized salt has you covered. If you are avoiding salt (not a bad idea), sea plants are a great source of iodine. Although most Westerners are new to eating sea plants, it's a good habit to cultivate. Next time you are at a sushi bar, have a seaweed salad or a vegetable roll wrapped in nori. If you get hooked, you can introduce these same tastes into your own kitchen. Try the Arame Salad, Rainbow Nori Rolls, or Kale & Sweet Potato Sushi in the recipe section.

There are two seaweed varieties that I would suggest avoiding: Kelp is *very* high in iodine—too high really. And hijiki can be contaminated with arsenic. Nori and wakame are better choices and are available at any health food store.

If you are pregnant or are planning to become pregnant, speak with your doctor about iodine supplements. Iodine is especially important for the cognitive development of a growing baby.

Avoid animal products. People who avoid meat, dairy products, and eggs have been shown in research studies to have the lowest risk of hypo- or hyperthyroidism. No one knows for sure why that is, but I suspect they have a measure of protection against the antibody attacks that can create havoc for the thyroid.

If you take thyroid medication, take it on an empty stomach. Many foods interfere with the absorption of thyroid medications. Also, it is a good practice to stick with the same medication brand over time, rather than switching from one brand to another. Each brand has a slightly different absorption, and if you stick with the same brand, you will find it is easier to achieve a consistent blood level.

Common thyroid diseases probably reflect a combination of genetic vulnerability and environmental triggers—with food being a prime "environmental" suspect. In other words, even if you have inherited a genetic vulnerability to thyroid problems, some food choices can cause problems to manifest while others will help you keep those problems at bay.

The good news is that, with appropriate attention to your diet—and medical attention when you need it—it is usually an easy matter to get your thyroid in good balance.

Steady as she goes, Mr. Sulu.

CHAPTER 10

Healthy Skin and Hair

Our skin and hair are strongly affected by hormones, as adolescent acne and male-pattern hair loss will attest. In turn, hormones are influenced by foods. In this chapter, I'll show you how you can put them to work for healthy skin and hair.

Healthy Skin

Acne often arrives as we make our way through adolescence. For some, it passes quickly. For others, it persists. And the perennial questions have been, are foods to blame, and, if so, which foods? Let's have a look.

But first, what is acne anyway?

A hair follicle is like a well with a tree growing out of it. The "well" is a pore in your skin and the "tree" is a hair. Normally, your natural oils pass upward from the "well" and out onto the skin surface. But sometimes oil production increases and the cells lining the sides of the "well" multiply, clogging the follicle, which then fills with bacteria.

Androgens—male sex hormones—encourage this process. Certain foods can, too. And sometimes foods and hormones conspire together. Let's take a closer look at the foods that have been under suspicion.

Chocolate

Many teenagers swear that chocolate triggers breakouts. So let's run a Google search for "Does chocolate cause acne?" and see what we come up with.

The top article comes from Verywell Health, saying in bold letters, "There Is No Evidence That Chocolate Itself Causes Acne. It's good news for all you chocoholics: eating chocolate does not cause pimples. There are no studies linking this sweet treat to the development of acne. There is no evidence that cocoa beans, from which chocolate is made, cause pimples."[1]

WebMD follows suit in an authoritative Q&A: "Q: Does chocolate really cause acne? My teenagers love the stuff—and they have pretty bad breakouts. A: Sorry, Mom and Dad. Your dire warnings about Snickers bars are fruitless, because the answer is FALSE. Chocolate has no link to acne (nor do other frequently blamed foods, such as pizza and potato chips)."[2]

The article backs up its assertion with science: " 'There was a famous experiment done many years ago at the University of Pennsylvania by Dr. Albert Kligman,' says Irwin Braverman, MD, professor of dermatology at Yale School of Medicine.

"Kligman gave teens with acne real chocolate bars, and others chocolate-free bars that tasted like chocolate. Neither group knew which candy bars were fake. 'The variation in the acne and induction of acne lesions was no greater in the chocolate group than in the nonchocolate group.' "

The Kligman study was quoted by many other Internet sites, too, as proof that chocolate has nothing to do with acne.

Wow, were all those teenagers wrong? Was the link between chocolate and acne all in their heads? Let's take a closer look. The

Kligman study was published in *JAMA* in 1969. As WebMD pointed out, Dr. Kligman and his colleagues gave research subjects either chocolate bars or look-alike bars made of partially hydrogenated fat. After a month, acne was about the same in the two groups.[3] Chock one up for chocolate.

However, the study was hotly controversial, for good reasons. First, while there were indeed some young people in the study, thirty-five of the sixty-five participants were actually inmates at Philadelphia's Holmesburg Prison at a time when prisoners did not have the protections they have today. Dr. Kligman was later vilified for a long series of experiments in which he exposed prisoners to things much worse than chocolate: He exposed them to dioxin (Agent Orange) in tests paid for by Dow Chemical, and also to infectious bacteria, viruses, and psychoactive drugs. His experiments were detailed in a book entitled *Acres of Skin: Human Experiments at Holmesburg Prison*, drawing its title from Kligman's comment on seeing the massive numbers of prisoners available to him for dermatologic experiments: "All I saw before me were acres of skin. . . . It was like a farmer seeing a fertile field for the first time." Although his chocolate tests were among his more benign studies, his experiments overall have been denounced as a classic example of unethical experimentation. In 1966, discrepancies in his data led the Food and Drug Administration to temporarily bar him from research.[4]

Second, the *JAMA* article says that part of the study "was made possible through the Chocolate Manufacturers Association of the U.S.A." Whether the financial relationship extended beyond providing the chocolate bars is not clear in the report.

Third, Dr. Kligman and his fellow researcher James Fulton had something to sell. They invented *Retin-A* (tretinoin), a commercial acne treatment. If a simple diet change could knock out acne, that would not be good for drug sales.

Other researchers have reexamined the chocolate-acne issue and have arrived at a very different conclusion. In 2016, researchers asked twenty-five acne-prone young men to have 25 grams (about an ounce) of Lindt 99 percent dark chocolate every day for four weeks. The ingredient list was simple: cocoa mass, cocoa powder, cocoa butter, and brown sugar. Unlike Kligman's experiments, chocolate did indeed trigger acne. Pimples showed up within two weeks and stuck around for the duration of the study.[5] Other researchers tested 100 percent chocolate and got the same result.[6]

In 2018, researchers looked to see what chocolate actually does to the skin. They asked a group of men to have dark chocolate after lunch each day—10 grams (about one-third of an ounce) of Green & Black's 70 percent dark chocolate from London. They then carefully collected the oils and sloughed cells on the skin surface. After four weeks, bacterial counts on the skin surface had noticeably increased and the number of sloughed cells had, too.[7] In other words, chocolate was indeed affecting the skin and promoting changes that can lead to acne.

Is it the sugar? Apparently not, because researchers have found the same effect with unsweetened cocoa—even when given in plain capsules that research participants simply swallowed.[8] Moreover, when researchers have compared chocolate and jelly beans, chocolate triggered pimples within forty-eight hours; jelly beans did not, suggesting that sugar is not the issue.[9] It is not clear what it is in chocolate that triggers breakouts.

By the way, chocolate's problems may not be dose-related. In other words, it's not as if a little chocolate causes a few pimples and a lot of chocolate causes a lot of pimples. It seems as if, in vulnerable people, just having chocolate at all—in any amount—can trigger acne.

We should acknowledge a vulnerability in all of these studies, which is that researchers love to find something new, something contrarian. So if everyone believes chocolate causes pimples, researchers love to disprove it. And when researchers exonerate chocolate, other researchers jump back in to condemn it. In other words, it pays to take the findings of any given study with a grain of salt. Even so, when one study after another produces the same result, the evidence becomes more convincing.

So, bottom line, there is actually good evidence that chocolate can cause acne. Whether that is true for you as an individual is another question. If you are locked up in a Philadelphia prison, a chocolate bar may not be any worse for you than a bar made of partially hydrogenated fat. But if chocolate has become part of your life and acne has, too, you may want to see if avoiding chocolate helps.

Dairy Products

Dairy products have been implicated in inflammatory conditions of all kinds, notably rheumatoid arthritis and asthma. Could they also inflame your skin?

In a 2005 study, Harvard researchers[10] asked more than 47,000 nurses about what they had eaten as teenagers and whether they had had acne. It turned out that those whose adolescent diets had included more milk were more likely to have had acne than those who generally avoided milk. Perhaps surprisingly, skim milk appeared to be a bigger problem than whole milk. The authors theorized that the hormones in milk products were responsible.

Because that study required participants to remember what they had eaten in the past—and their memories might have been less than accurate—the researchers did a new study. This time, they followed the eating habits of 4,273 boys and 6,094 girls over time, to see if the foods they ate were linked to acne outbreaks.[11,12] Again, it turned out that youngsters having two or more milk servings per day were about 20 percent more likely to have acne, compared to those who mostly avoided milk.

There are plenty of reasons why dairy products could affect your skin:

First, dairy products contain hormone traces that came from the cow, as we saw in Chapter 1, "Foods for Fertility." In addition to estradiol, milk contains 5α-pregnanedione and 5α-androstanedione, both of which can be converted in hair follicles to dihydrotestosterone (DHT), which is thought to be the main acne trigger.[13]

Second, milk boosts IGF-1 levels in your body. IGF-1 is short for *insulin-like growth factor*, which we briefly mentioned in Chapter 4, "Tackling Cancer for Men," because of its relationship with prostate cancer. IGF-1 stimulates the oil-producing cells in your follicles.[14]

Third, dairy proteins can trigger sensitivities of all kinds, as some people with asthma and joint pains have discovered.

Fourth, most dairy products are loaded with saturated fat.

For acne and any other sort of inflammatory condition, it makes sense to see if avoiding dairy products helps. To do a good test, you'll want to avoid dairy products *completely*, including nonfat milk and foods that include milk proteins (e.g., casein) as an ingredient. The protein in milk may be as problematic as the fat.

A Look at the Whole Plate

Maybe the problem goes beyond chocolate and milk. Maybe the problem is a not-so-healthy diet overall. Some researchers have pointed a finger of blame at meaty, dairy-laden Western diets. Others have focused on processed foods, wheat flour, white potatoes, table sugar, and breakfast cereals.

Some evidence comes from rural societies eating simple, mostly unrefined, dairy-free diets. Among the Kitavan islanders of Papua New Guinea, for example, the traditional diet was based on root vegetables, fruit, fish, and coconut. Their intake of sugar, coffee, alcohol, cereals, oils, and salt was minimal. Similarly, among the Aché of Paraguay, most of the diet consisted of sweet manioc, peanuts, maize, and rice, with much smaller amounts of flour, sugar, and meat. There were no reports of acne in either population.[15]

As cultures change and populations Westernize, acne starts to appear. Health reports of northern Canadian Inuits made no mention of acne before soda, processed foods, beef, and dairy products arrived from their southern neighbors.[16] Similarly, as the traditional Okinawan diet emphasizing sweet potatoes, rice, and vegetables, along with some soybeans, but little meat, gave way to a meatier, more processed, Westernized diet, acne became more common.[17]

Putting the theory to the test, Australian researchers fed volunteers foods that were less refined, with a lower Glycemic Index. As you will recall from Chapter 8, "Conquering Diabetes," the GI separates foods that raise blood sugar quickly (e.g., white bread) from those that are more gentle on blood sugar (e.g., rye bread, fruit, beans, and pasta—yes, surprisingly, pasta has a low GI because it is compact and dense, so it digests slowly, unlike white bread). Over a twelve-week period, the volunteers had fewer and fewer acne lesions.[18]

So it looks like simpler, more natural diets that are free of chocolate and dairy products and are less processed overall may be helpful.

Nina and Randa

Nina and Randa are identical twins. During adolescence, they had occasional pimples, though nothing serious. But at age twenty, for some reason acne arrived with a vengeance, and it became increasingly severe.

This was not good. They were young actors and musicians, and their stage appearance mattered. Even in day-to-day interactions with their friends, they were embarrassed by acne covering their faces.

A doctor prescribed antibiotics, which were helpful at first. But when the prescription was done, the acne came right back. So they repeated the antibiotics, and with each medication course, the pills were less and less effective. Their acne ended up worse than ever. They consulted another dermatologist—this one very well reputed—who recommended Accutane (a vitamin A derivative, whose generic name is isotretinoin). However, friends who had taken the drug had had terrible side effects—depression, hair loss, and colitis. Indeed, it has an intimidating list of adverse effects. Visits to two other dermatologists led to the same recommendation: Take Accutane and hope for the best. They decided against it. There had to be a better way.

Acne, they discovered, is not just a physical problem. "Acne makes you depressed," Randa said. "Some people say it's not a big deal; it will go away. But when you have painful zits, it affects your self-concept, your personal life, your career—everything."

"We had a hunch it was something we were eating," Nina said. "But we could not figure out what it was. We never ate chocolate anyway. We tried avoiding high–Glycemic Index foods and limiting carbohydrates." Nothing worked.

Then, while driving to visit their grandparents, their father suggested checking Dr. John McDougall's website. Dr. McDougall, a well-known physician and author in Santa Rosa, California, has helped thousands of people using a nutritional approach for serious and not-so-serious conditions. He had helped their mother overcome a major health problem; perhaps he would have some advice for acne. They pulled up his website and indeed, he had some recommendations. First, avoid animal products, something they were doing already. In fact, they had been raised on a vegan diet. He went a step further and recommended keeping fats in general to a bare minimum. The idea was that even small traces of fatty foods could trigger acne in a susceptible person.

They decided to give it a try. Out with the peanut butter, guacamole, and other fatty foods. To their surprise, the results came almost immediately. Within a few days, new pimples stopped arriving, and old pimples started to heal. As the weeks went by, their acne gradually disappeared. Now four years and counting, it has never returned.

Impressive. But was it hard to do? "Actually, it was extremely easy for us," Randa said. "When we make oatmeal, we cook it with just water. We love brown rice, sweet potatoes, broccoli, spinach, corn, beans and rice, applesauce, and pasta with spices of all kinds. We have lots of fruit—fresh, dried, or in smoothies with oats." Eating out means Asian restaurants, especially Thai, for noodle and vegetable dishes or steamed rice. The twins recently visited a steakhouse and got potatoes and vegetables. They bought an Instant Pot, which is a pressure cooker that whips up meals in a jiffy. And for fun, they have a Yonanas machine, which makes vegan soft serve from bananas and other healthful foods. They love the food, the fun of trying new things, and the healthy skin a diet adjustment has brought them.

Nina and Randa wrote *The Clear Skin Diet* and set up a website, ClearSkinDiet.com, both focused on a healthy low-fat, vegan diet. They offer additional tips, too:

1. Have a friend or family member join you as you change your diet. It turns the process into an adventure.
2. Give it time. Recovery can take awhile—maybe months in severe cases. So don't stress. Let foods do the work for you.
3. If you do not see the results you are looking for, see if other foods might be a problem: especially sugar, gluten, and processed foods. Choose oil-free skincare products (moisturizers, sunscreen, makeup), shampoo, and conditioner.

PCOS and Acne

As we saw in Chapter 5, "Reversing Polycystic Ovary Syndrome," acne can also be a sign of polycystic ovary syndrome. The problem, in this case, is an excess of androgens—male sex hormones. Along with acne, women with PCOS may also have irregular periods, thinning hair, unwanted facial or body hair, and, as its name suggests, ovarian cysts. Doctors can diagnose and treat this condition, and a key part of the treatment is a healthful, plant-based diet.

Healthy Hair

Can foods help you keep a healthy head of hair? It is a fascinating area of research—and one that is by no means finished. Let me share what we know.

Common age-related hair loss is influenced by genes, of course. In some families, baldness runs up and down the family tree. But *the process is entirely dependent on hormones*, as was dramatically demonstrated by Yale University's James B. Hamilton in 1942. Hamilton showed that men who, for whatever reason, had been castrated before puberty never lost their hair. Even if every last male in their family had gone bald, it did not happen to them. Without testosterone, the genes for hair loss did not express themselves.

Hamilton also found that when men who were losing their hair were castrated (for medical reasons having nothing to do with hair loss), their hair loss suddenly stopped. If castrated men were then given supplemental testosterone, hair loss kicked in. If the testosterone treatments were stopped, hair loss stopped, too.[19]

Here is what is going on: In the hair follicles, testosterone is converted to *dihydrotestosterone* (DHT). DHT is the baldness trigger. In genetically susceptible individuals, it causes hair follicles to gradually shrink the size (length and diameter) of the hairs they produce until finally they stop producing hair altogether.

Some parts of the scalp—especially the front and crown—are particularly sensitive to DHT, while others—the sides and back of the head—resist it. And on the face and chest, DHT has the opposite effect. It stimulates follicles to produce thick, curly hair.

The conversion of testosterone to DHT can be blocked by *finasteride*, a drug marketed under the brand name *Propecia*. Rogaine (minoxidil) works differently. It is used topically to keep follicles functioning normally.

Foods and Hair Retention

So, how do foods fit into this? For starters, researchers noticed that baldness was less common among Asians than in whites.[20] But as Asian countries began to Westernize their diets, baldness was one of the conditions that appeared to be increasingly common. Japan is a case in point. As its diet became Westernized in the latter half of the twentieth century, many aspects of health changed, as we have seen. Although most of the attention went to the massive rise in breast cancer, diabetes, and heart disease, dermatologists also noted that baldness became more common.[21] The same phenomenon was observed in Korea. Doctors in a dermatology clinic found that baldness seemed to be striking early and more often.[22]

A traditional Asian diet is based mainly on plant-based foods, rather than meats and dairy products, and tends to be very modest in fat. As we saw in Chapter 1, a low-fat, plant-based diet encourages your body to build more SHBG—sex hormone–binding globulin. SHBG reins in testosterone and keeps it inactive until it is needed. That's good. You will have more than enough testosterone for your daily needs, without the excesses that could affect your scalp.

One more thing: In 2009, researchers compared eighty young men with progressive hair loss to eighty men without hair loss. The balding men were more likely to have insulin resistance. Other studies found the same thing: Insulin resistance is linked to hair loss in both men and women.[23,24] This means that the cells of the body (especially the muscles and liver) have become unresponsive to insulin, as a result of the buildup of fat inside the cells, as we saw in Chapter 8. In turn, insulin resistance causes metabolic changes that affect your whole body. It can impair blood circulation to the follicles and contribute to the loss of the follicles' ability to produce hair.

This can happen to women, too. Women with polycystic ovary syndrome, in particular, have higher levels of testosterone than usual and often have insulin resistance. Many have thinning hair.

There are other causes of hair loss, too. Thyroid disease and various medications can cause hair loss.

Keeping Healthy Skin and Hair

To maintain healthy skin and hair, I would encourage you to:

1. Avoid animal products.
2. Avoid adding oils in cooking, and favor oil-free foods at restaurants.
3. Avoid oily foods (e.g., peanut butter, avocados) until you know how they affect you.
4. Avoid added sugars. In anecdotal reports, for whatever reason, sugary foods seem to make hair lifeless.
5. Have adequate protein *from plant sources.* Beans and bean products, such as tofu, tempeh, and soymilk, give you plenty of protein without the negatives of dairy products or meat. Anecdotally, some people have found that having extra plant protein helps keep their hair fuller.
6. Have plenty of vegetables and fruits. Their antioxidants will help protect your skin.
7. Protect your skin from excess sun exposure.
8. Although hormone shifts can have profound effects on your skin and hair, a healthful diet can be powerful, too. See what healthful foods can do for how you look, in addition to how you feel.

CHAPTER 11

Foods That Fight Moodiness and Stress

In Chapter 1, "Foods for Fertility," I described how simple diet changes can help you conquer the moodiness of PMS. Evidence suggests that this is also true for more serious mood problems, particularly depression and anxiety. For some people, a switch to healthier food choices is like day and night.

Imagine what this could mean. If you have felt that you are not at your best and that you just have to put up with it, we may have a better answer. If bouts of depression or anxiety have had you in their grip, rethinking what's on your plate could be a key part of the solution.

First, a note of caution: Depression is dangerous. In the depths of despair, suicide looms as a real risk. Medications used to treat it have risks, too. Also, depression can have many causes and complicating factors. It is important to get professional help and to use the information in this chapter as part of your treatment program, not in place of it.

Joy

Joy is a schoolteacher in New Jersey who discovered some remarkable effects of foods. For quite some time, she had not been eating especially well. Fast food, fried chicken, and steak figured in her routine, as did cheese. She loved a block of cheddar with crackers, shredded cheese on salad, and Tex-Mex recipes at home or at fast-food restaurants.

Over time, her weight gradually climbed from 145 to 190 pounds. Her cholesterol level rose, too, and she had frequent upper respiratory infections. She never really felt well. Her energy was often low, and a predictable midday slump left her ready for a nap.

Most troubling of all was a persistently low mood. She often felt depressed, irritable, and anxious, and had troubling mood swings. Before her periods, she teared up easily, and her cramps were terrible.

She had had her share of traumas: a divorce after three years of marriage, with three miscarriages during those three years, and the death of both her parents soon thereafter. Because of her miscarriages, she found it hard to see women in late pregnancy or mothers with their small children; she wondered if she would ever be able to be a mother herself. She expected that she should eventually be able to recover her equilibrium. But that didn't happen. Her low moods would not go away.

One day, at the library, she spotted a book of mine called *The Cheese Trap*, all about the health effects of cheese. As you know by now, cheese contains hormone traces, not to mention a lot of fat, cholesterol, and calories. Hormones have physical effects, of course. But they can also affect us emotionally—how we feel from day to day. That is true for the hormones our bodies make. Could it be that the hormone traces found in food can affect our moods, too? As she paged through the book, the symptoms I had discussed sounded very familiar.

After considerable reflection, she decided she needed to change her eating habits. Out with the cheese, she decided. In fact, out with dairy products altogether. She jumped into a healthier eating plan and stuck with it.

Over the next few months, she felt better and better. Upper respiratory infections stopped, her cholesterol level dropped, and her cramps were dramatically reduced.

Within three months, her PMS was gone. There was no midday slump at work, she slept deeply through the night, and she had better mental clarity. Episodes of depression or anxiety were fewer and milder. She felt a stability and a vitality that had eluded her for years.

Things were not perfect. She still felt overly sensitive at times. Even so, she was less inclined to wallow in hurt feelings. "The stability in my moods was a great and welcome relief," she said. "I felt more balanced."

Unfortunately, celebrations at work conspired against her. In came the macaroni and cheese and all the other not-so-healthy foods, and eventually she slipped off the wagon. Returning to her previous eating habits, she soon was right back where she had been. Her cramps returned. Allergies worsened, and her anxiety and depression returned.

And that did it. She realized that foods really were affecting her mood. She decided to reclaim the sense of balance that came with a healthier diet and began transitioning back to the foods that helped her feel good.

Kim

Kim is an interpreter, living in Los Angeles with her husband. As a child, her health was mostly good, apart from having a digestive tract that never seemed to work right—she was plagued by persistent constipation. Otherwise, things were more or less normal.

At age fifteen, for no apparent reason, her mood took a nosedive. Out of nowhere, she found herself crying, lost her appetite, and sank into a depression. The normal pleasures of life were gone. Her mother took her to a counselor who recommended tryptophan supplements, which she dutifully took.

Things worsened. At age eighteen, anxiety and indecisiveness took over. Even simple decisions, like which clothes to wear, left her helpless, and her depression reached the point where she found it hard to leave her home. This was not good, especially since she was enrolled in classes at a community college thirty-five miles away. Sometimes, she drove to the campus, only to be overcome with sadness to the point where she could not get out of her car. She turned around, drove home, and sat in the corner with a pillow and cried. At one point, driving back from her college, she unbuckled her seat belt and blew through a stop sign without stopping, hoping someone would hit her and kill her.

She tried psychotherapy, but it did nothing. It was as if her problem was *physical*, not really related to her life circumstances at all.

A psychiatrist started her on a series of antidepressant medications, which she continued for the next nineteen years. They helped—sort of. Between the depression and her medications, she mostly felt numb. She tried stopping the medications, because she did not like the way they made her feel, but her symptoms soon returned and she resumed the medicines. It seemed like nothing would ever get better.

One day, she heard a lecture from a psychopharmacologist who said, "Your gut makes more serotonin than your brain." This rang a bell with her. If the digestive tract makes neurotransmitters, maybe her chronic digestive issues were somehow affecting her brain chemistry. Maybe a diet change could help.

Up until that point, her favorite foods had been boxed macaroni-and-cheese dinners, fried chicken, hamburgers, mashed potatoes and gravy, biscuits and gravy, and chicken wings with Buffalo sauce. She decided to try something new.

She started taking a probiotic and, over time, added more vegetables and fruits to her diet. She also switched from cow's milk to soymilk and eventually stopped eating meat. With these changes, she began to feel better. Although she still had episodes of what she called "doom and gloom," they were not as severe as before. She was not really well, but she was better.

Three years later, she took another step. She decided to go vegan. No more animal products at all—no cheese, no yogurt, nothing, unless they were plant-derived. Her new favorites were beans of all kinds—black beans, garbanzos, pintos, and cannellini beans—along with potatoes, big salads with kale, vegetables, mandarin oranges, and nuts, plus bananas, pineapples, and other fruits.

And at that point, everything changed. Her long-standing digestive symptoms disappeared. Even better, the mood symptoms that had been all but disabling for so long gradually just faded away.

It wasn't just the chronic depression that lifted. The same was true for the monthly moodiness of PMS. "I had always suffered from horrible mood swings as a result of PMS," Kim said. "So much so that my husband used to joke that he wanted the earth to open up and swallow him for the entire week before my cycle began. That changed almost immediately. In fact, the first month after I switched over to an all-vegan diet, I had absolutely no symptoms of PMS, and was actually surprised when my period arrived. The change was incredible, and to this day, I don't suffer PMS like I did for my entire life up to that point.

"As I healed my digestive system," Kim said, "my moods improved, too." She has now been off all medications for thirteen years, and she no longer suffers from depression. In the bargain, chronic allergies—to mold, pollen, and many other things—all disappeared as well.

Today, she asks, why do researchers not focus on the role of foods in mental health? We treat depression with drugs and psychotherapy. Although they certainly have an important—even lifesaving—role, a fresh look at the interplay between nutrition and the brain may lead us to a new understanding of what controls our emotions and better ways to address them.

Inside the Brain

Depression does not mean being down in the dumps for a day or two. Depression is a dark and oppressive cloud that continues for weeks on end.

Depression is also not the same as the sadness that comes from suffering a loss or getting bad news. These stressors sometimes lead to depression, but often the grief of bereavement is very different, coming in waves that are punctuated by memories—sometimes happy memories—of the person you've lost. Eventually, the sadness lifts on its own. In depression, the low mood is not focused on a particular person or event, and it lingers on.

In diagnosing *major depressive disorder*, psychiatrists first look for persistent depressed mood or a loss of interests. They also look for changes in sleep and appetite. Usually, you'll have insomnia and a poor appetite, but sometimes the opposite can happen: You're eating more than usual and having trouble dragging yourself out of bed. Along with these changes come restlessness or fatigue, guilt, diminished concentration, and thoughts of death. The symptoms vary from person to person.

Although stresses play an important role in moods, depression can also be triggered by *purely physical factors*. We were taught this by reserpine, a blood pressure medicine developed in the 1950s. Patients taking reserpine sometimes fell into inexplicable depressions. People who had never been seriously depressed before suddenly just ran out of gas, lapsing into depressions that were sometimes so serious they had to be hospitalized.

The explanation came from how the drug works. Reserpine reduces the effects of neurotransmitters that control blood pressure; as these brain chemicals diminish, blood pressure comes down. However, these same neurotransmitters control your emotional state. Losing them is like pulling the plug on your mood.

It soon became clear that depression is a side effect of several medications. These include Premarin—the horse-derived estrogen medication we discussed in Chapter 6, "Tackling Menopause"—as well as steroids, cholesterol-lowering drugs, and many others. Depression can also result from alcohol, opioids, amphetamine, or cocaine.

Physical illnesses cause it, too. When I was a medical student, I was surprised to see the deep depressions that could follow a stroke. Not uncommonly, depression follows heart attacks, childbirth, and surgical procedures of all kinds—even gastric bypass surgery, cosmetic surgery, and hip replacements that are supposed to make you feel better. It's as if the body is concentrating on healing its injury and does not have the ability to maintain a good mood at the same time.

Foods That Fight Depression

If medicines and physical illnesses can influence your mood, it should be no surprise to find that foods can, too. In a 2010 study, researchers compared 60 vegetarians to 78 omnivores, giving them questionnaires that assessed their mood. They found that the vegetarians scored better.[1] A larger study, this one including 620 people, found more or less the same thing.[2] For some reason, people following vegan diets reported less anxiety and stress, compared with omnivores or lacto-ovo vegetarians.

One might have thought the opposite. After all, people who have gone vegan have to deal with family members and friends pestering them with questions like, "Where do you get your protein?" But, no. The people on vegan diets actually had better moods.

Similarly, a large study in Spain found that people eating very little meat were less likely to be hit by depression, compared with those who ate more meat.[3] Following their participants over a total of eight years, the researchers found that diets built mainly from plant sources were associated with a 26 percent reduction in the risk of depression.[4]

An Antidepressant Effect of Fruits and Vegetables?

It is apparently not just a question of what you *avoid*. The foods you *include* seem to matter, too. In the Spanish study, there seemed to be a particular benefit from fruits, nuts, and beans. People eating these foods were less likely to lapse into depression. Similar findings came from a meta-analysis of four prior studies examining a "Mediterranean" diet pattern. It was associated with about one-third less risk of depression, compared with more omnivorous diets.[5] Similarly, a large study in Taiwan examined the diets of 1,609 older men and women, finding that those who ate the most vegetables were 62 percent less likely to develop depressive symptoms, compared to their vegetable-neglecting friends. Fruits also helped.[6] In a study of approximately 50,000 individuals in the United Kingdom, those consuming more vegetables and fruits reported substantially better mental well-being and life satisfaction, compared with their friends who neglected these healthful foods. And the more vegetables and fruits they ate, the better they did on these measures.[7]

Other studies have shown similar results: There is something about a plant-rich diet that seems to improve mood and reduce the risk of falling into depression.

Is Dairy Depressing?

Dairy products contain *casomorphins*—mild opiates that are released as milk is digested. *Casein* is the main protein in milk. If you could look at it under a powerful microscope, it would look like a string of beads—each "bead" is an amino acid. As casein breaks down in your digestive tract, individual "beads" are released into your bloodstream. But casein also releases short strings of "beads"—each string is made up of just a few amino acids. They, too, pass into your blood and reach the brain, where they attach to the same receptors that heroin or morphine attach to. These are casomorphins.

That's right. Milk contains traces of opiates. Presumably, Mother Nature built these traces into cow's milk as a way of calming the calf. However, in adult humans, they may have some untoward effects. For one thing, as milk is turned into cheese, these opiates become more concentrated and may be one of the reasons that cheese is habit-forming for many people. Like other opiates, casomorphins can be constipating, as you may have noticed if you have ever overdone it on cheese.

Swedish researchers discovered that, when highly concentrated, casomorphins can have a dramatic effect on the brain. They examined women with postpartum psychosis, a severe psychiatric disorder that affects about 1 in 1,000 women after giving birth. It starts with insomnia, restlessness, irritability, and depression, soon leading to delusions and hallucinations. Surprisingly, the researchers found casomorphins *in the cerebrospinal fluid* (the fluid that surrounds the brain and spinal cord) of many of these women.

Because the women had all given birth, they had begun producing breast milk. The researchers believe that, within their breasts, casein molecules had broken apart, releasing casomorphins into their bloodstreams. From there, the casomorphins reached their brains, causing massive derangements of brain function.[8,9] Just as a small dose of heroin would cause a user to feel relaxed and a larger dose could bring on strongly negative and even life-threatening symptoms, the same may be true for the weaker opiates found in milk and especially in cheese.

The Swedish researchers found that, in unusual circumstances, a woman's own breast milk can release a hefty dose of casomorphins into the blood. If you consume dairy products, especially cheese, you are getting casomorphins—from a postpartum cow—in every mouthful.

Intriguing as these findings are, I believe we need more research to understand the effects of dairy products and casomorphins on the brain. However, the role of these compounds in psychiatric disorders is a fascinating evolving story. The women mentioned at the outset of this chapter had noted that breaking a cheese habit made them feel dramatically better, and others have noted the same thing, too.

A Mood-Boosting Effect of a Plant-Based Diet

A plant-based diet appears to have a mood-stabilizing effect for many people. You may wish to see if the same is true for you.

A team of researchers decided to put a plant-based diet to the test.[10] Randomly assigning volunteers to a diet that included meat, poultry, and fish; a diet with no meat other than fish; or a diet that included no meat of any kind, the researchers found that the meaty diet and a fish-based diet did nothing for mood, but the vegetarian diet significantly boosted mood.

These results resonate with those of my own research team. In the course of two large studies, we looked at how foods affect depression and anxiety.[11,12,13,14] The studies took place at GEICO, the car insurance company, in ten different cities around the United States, including Dallas, Texas; Macon, Georgia; Buffalo, New York; San Diego, California; and elsewhere. The overall goal of the studies was to see whether a diet change could help people lose weight and improve their health at work.

The program we were testing involved two components. First, employees participated in weekly lunchtime classes that helped them learn about a healthy vegan diet. We had lectures, cooking demonstrations, and plenty of time for discussion. In addition, the company cafeteria served healthful foods every day: veggie burgers, vegetable curry, black bean chili, veggie lasagna, and other tasty choices.

As the weeks went by, we checked everyone's weight, cholesterol level, blood pressure, and other measures. The results were exactly what you would expect: Their health had improved substantially. However, we also zeroed in on our participants' moods, using a structured psychological questionnaire. We found that as people changed their eating habits, depression and anxiety remitted to a significant degree. The changes were not necessarily dramatic; they just felt better. Work productivity improved, too, and absenteeism fell.

Dr. Dean Ornish found the same thing. We touched on Dr. Ornish's program for reversing heart disease in Chapter 7, "Curing Erectile Dysfunction and Saving Your Life." The program includes a plant-based diet, along with other healthful lifestyle changes and has become famous for its ability to reopen narrowed arteries and to give heart patients a new lease on life. Dr. Ornish developed a similar program for prostate cancer patients, as we saw in Chapter 4, "Tackling Cancer for Men." A careful study of nearly 3,000 program participants in twenty-four different hospitals not only showed major physical improvements; psychological testing showed dramatic improvements in depression scores. The improvement was clear within twelve weeks and continued over the next year.[15] In the bargain, hostility scores improved, too, so participants were better able to cope with whatever life might bring.

Partly, these results could be explained by the fact that the research participants were losing weight and regaining their health. Those changes alone would be energizing for anyone. But there is more to it. Healthy foods can change brain chemistry in such a way as to get our moods on a more even keel.

A low-carbohydrate diet does not seem to have the same effect, according to a research study in Australia. As you know, diets eliminating bread, potatoes, and other carbohydrate-rich foods are a recurring fad, despite their not-so-healthful effects on cholesterol levels, digestion, and other aspects of health. So researchers asked 106 overweight volunteers to follow either a low-fat diet or a low-carbohydrate diet in a one-year study. The two groups lost similar amounts of weight, but the two diets had very different effects on mood. For the first eight weeks, both groups seemed to be doing well, losing weight and feeling better overall. However, after that point, things changed. While the low-fat group continued to show mood improvements, the low-carbohydrate group worsened, showing considerably more depression, anxiety, and hostility, compared with the low-fat group.[16] The same researchers then did a new study in which they modified the low-carbohydrate diet, boosting carbohydrates a bit, reducing the saturated ("bad") fat content, and adding exercise, and found that those changes seemed to help the mood scores in the participants.[17]

Westernization and Depression in Japan

The apparent antidepressant effect of a healthy diet could help explain a phenomenon in Japan. As you'll recall, hot flushes and hormone-related cancers were rare in Japan forty years ago, when the diet was based mainly on rice and vegetables. But as this mostly plant-based diet gave way to burgers, chicken, and cheese in the 1980s and 1990s, menopausal symptoms and cancer came flooding in.

(continued)

A strikingly similar observation was made for mood problems. A 2016 BBC news story described how depression was rarely reported in Japan until the 1990s.[18] Indeed, Asian countries in general have reported a lower prevalence of depression, compared with the United States. When cases began to crop up, individuals had no idea what was happening to them. Indeed, a computer search for depression in Japan would turn up articles about the country's *economic* depression in the 1920s, but nothing about psychological depression prior to the Westernization of the diet. Today, depression is widely reported in Japan, and the pharmaceutical industry has done its best to capitalize on it.

One might attribute this to stoicism or a sense of shame on the part of Japanese people in decades past—that is, depression may have existed but was simply not talked about. If so, it is not clear that that reluctance would have changed over time. Or maybe the apparent rise in depression is due to the notoriously workaholic lifestyle in Japan that sacrifices individuals' social lives. Or were the Japanese always ready to work from early morning to late at night?

Certainly, Japanese society has changed in many respects. In a culture where, traditionally, dairy products were virtually never consumed, meat was a modest part of the diet, and fatty foods in general were scarce, depression has paralleled the influx of unhealthful foods. Patient surveys indicate that depression became much more common with the approach of the new millennium, as dairy intake was reaching unprecedented levels, meat intake was continuing to rise, and rice consumption had fallen to half its pre–World War II levels.

Anti-Inflammatory Foods

How could shifting to a plant-based diet be a mood-booster? One way may relate to *inflammation*. Let me explain:

When you are stung by a bee, your skin turns red and swells. The same thing happens when you cut your hand. Your injury turns red, swells up, and becomes warm to the touch. That's inflammation.

When your house is burning, a fire station sends trucks to put out the fire. When you are injured, your body sends specialized protein molecules to the site of injury. They widen the blood vessels in the area to increase blood flow and weep fluid and proteins into the injured area. That's where the redness and swelling come from. Your body also sends in white blood cells that produce antibodies, microscopic torpedoes that attack bacteria and viruses. They neutralize and engulf these invaders, destroying them before they can harm you.

Fire alarms can be triggered by mistake, and that's true of inflammation, too. Even when you have not been hurt, your body can make inflammatory proteins and antibodies and release them into the bloodstream. They can affect your joints, for example. In rheumatoid arthritis, antibodies are mistakenly attacking the synovial lining of the joints, leading to inflammation and pain. This process can be triggered by food. Dairy products are most notorious in this regard, but many other triggers have been identified, too.

A diet change can be like the water that puts out the fire. Many years ago, studies showed that avoiding animal products could reduce the inflammation of rheumatoid arthritis.[19,20] To prove that foods really can tackle inflammation, researchers measure *C-reactive protein* (CRP), a marker of inflammation. A vegan diet has been shown to lower CRP levels by about one-third.[21]

Inflammation affects more than your joints. When tennis star Venus Williams developed *Sjögren's syndrome*, an autoimmune inflammatory condition causing fatigue, chronic pain, and other symptoms, she threw out all animal products and used an entirely plant-based diet to tackle the disease and get her career back on track.

Here is the point: Inflammation can affect all of you, including your brain. No, your brain will not swell, turn red, or get warm, like a beesting. But inflammation can cause chemical changes in the brain. Years ago, blood tests showed that people suffering from depression had higher levels of inflammatory proteins in their bloodstream, compared with people who were not depressed.[22] After a considerable amount of research, scientists have come to view the chemical changes of inflammation as a major contributor to many cases of depression.[23,24]

More worrisome is the fact that inflammation may actually damage the brain. In other words, it does not just impair the brain cells' ability to function. It may be contributing to their destruction. Studies show that the more frequent and severe the depressive episodes a person suffers, the more brain cell destruction there is. Not surprisingly, depression is linked to a higher risk of dementia.

These observations suggest that just as inflammatory foods can affect the body, they may also affect the brain. And that suggests an explanation for the antidepressant effect of plant-based diets. They allow you to steer clear of inflammatory foods.

Apart from food triggers, the type of fat in the diet may make a difference, too. Researchers have suggested that an excess of *arachidonic acid*, an inflammatory fat found in chicken, eggs, beef, sausage, and fish, can promote inflammation in the brain and interfere with normal brain function.[25] The Australian researchers described above noted that the antidepressant effect of their low-fat diet might have related to the fact that it eliminated arachidonic acid.

Stay tuned. For now, we know that plant-based foods are good for the brain; the explanation as to why they help may become clearer with time.

Soy Joy

Soy products—tofu, tempeh, and soymilk, for example—contain *isoflavones*, which have been studied for their mood-stabilizing effect, among many other health benefits. While Americans tend to eat only modest amounts of these foods, they are staples in many Asian countries. While evidence is mixed, studies suggest that consuming two to four soy servings per day may bring a significant antidepressant effect.[26] This may be one possible explanation for the low prevalence of depression among Japanese people on traditional diets, and the increasing incidence of depression as their traditional diet is abandoned in favor of meaty, cheesy, Western foods.

A Healthier Digestive Tract

Plant foods provide fiber, which gives you a healthier digestive tract. In turn, that seems to protect against depression, surprisingly enough.

The National Health and Nutrition Examination Survey (NHANES) of 2007 to 2014 was an enormous survey that linked Americans' eating habits and their health. The NHANES data showed that the more fiber people ate, the less likely they were to be depressed.

Here are the numbers: The average American currently gets about 16 grams of fiber each day, about half the amount they need for good health. In the NHANES study, those who got around 21 grams per day were 41 percent less likely to report symptoms of depression, and at higher fiber intakes, the risk of depression dropped as much as 70 percent.[27]

Well, that's wild, isn't it? Of course, fiber fights constipation. So a person on a high-fiber diet has to feel better than someone who hasn't gone to the bathroom for a few days! But high-fiber foods

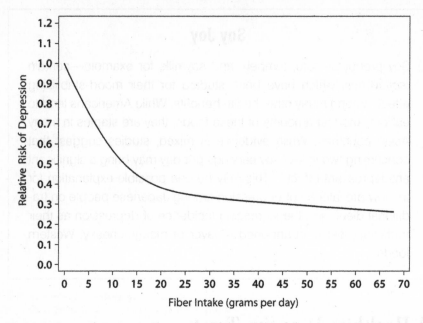

Xu H, Li S, Song X, Li Z, Zhang D. Exploration of the association between dietary fiber intake and depressive symptoms in adults. *Nutrition*. 2018;54: 48–53.

do more than that. Fiber stabilizes blood sugar. And as fiber ferments in the digestive tract, it releases short-chain fatty acids that may also have favorable brain effects.

Perhaps most important of all, fiber fosters healthy gut bacteria. When your intestinal bacteria are disrupted, either by an unhealthful diet or the use of antibiotics, depression is more likely to occur.[28] In the same way that the right kind of soil helps your garden grow, having plenty of fiber in your digestive tract helps healthy bacteria to flourish and keeps unhealthy bacteria in check. In turn, healthy gut bacteria appear to protect against depression.

So, where do you find it? Fiber is roughage from plants. There is lots of it in beans, vegetables, fruits, and whole grains, but none at all in meat, dairy products, or eggs.

In explaining the apparent antidepressant effect of plant-based foods, researchers have also credited folate, a B vitamin found in broccoli, spinach, asparagus, and many other vegetables, as well as in beans, peas, lentils, and chickpeas. Folate appears to play a role in the synthesis of serotonin, a neurotransmitter involved in moods.[29] They also suspect that the fact that plants have almost no saturated fat and proportionately more omega-3 ("good") fat, compared with meats, helps keep brain function on a more even keel. But the fact is, we are not sure which of these mechanisms is most important. It may well be that foods calm our hormone storms, reduce inflammation, and reestablish our gut health all at the same time. In any case, a healthful diet is well worth it, and all the side effects are good ones.

Plant Protein

We saw in Chapter 2 that a serving of plant protein as you start your breakfast can help stabilize moods for the rest of the day. At least that is what many women have reported. Although this has been suggested as a treatment for PMS, it also makes sense to put it to work for depression.

Many cultures have high-protein plant-based foods for breakfast: beans on toast in England, black beans in Mexico, and hummus in the Middle East.

I would suggest starting your breakfast with grilled tempeh, scrambled tofu, veggie bacon, veggie sausage, soymilk, or a small serving of beans or chickpeas. You do not need much. Starchier foods—toast, oatmeal, bagels—are fine, but have the plant-protein food first.

This does not mean that you are protein-deficient. A vegan diet brings you plenty of protein. We are just taking advantage of a mood-stabilizing effect that seems to come from plant proteins consumed early in a meal. Remember: We are speaking of *plant-based* proteins, not bacon, sausage, eggs, and so on. Animal-based foods cause the very problems we are trying to fix.

Vitamin B$_{12}$: Essential for Brain Health

Vitamin B$_{12}$ is essential for healthy brain function and also for healthy blood cells, and may play a role in regulating your moods.[30] You don't need a lot of it; adults need just 2.4 mcg per day, but you do not want to run low. For anyone on a vegan diet (as I heartily recommend), a B$_{12}$ supplement is essential. It is highly recommended for everyone else, too. Please see Chapter 12, "A Healthy Diet," for details.

Get Your Heart Beating

So far, we have focused on food choices. But your brain is also affected by physical activity of all kinds. When you are under a dark cloud of depression, you may not feel like going out and working up a sweat. But if you can bring yourself to do it, you will be rewarded. For many people, exercise counters depression.

Researchers at Duke University in North Carolina enrolled 156 depressed individuals, aged fifty and over, into a research study, assigning them to antidepressant medication, exercise, or both. The medication was sertraline, the drug marketed under the brand name *Zoloft*. It turned out that all the treatments worked. The medication worked a little more rapidly, but by sixteen weeks, they all had about the same effect.[31] The study suggested that exercise was as good as medication.

However, because the exercise was done as a group, it raised the possibility that it was actually the group support, not exercise, that deserved the credit. So researchers in Dallas, Texas, began a new study in which young, depressed people began a program of exercise one-by-one in private, using a treadmill or stationary bicycle, with no group support at all.[32,33] Some were asked to exercise three times per week for an hour. That would mean walking or jogging for about four miles each time, or the equivalent amount of exercise on a bike. Others were asked to do a bit less—around eighty minutes of exercise per week. After twelve weeks, everyone filled out a standardized mood questionnaire. It turned out that those with the more modest exercise program improved 30 percent, on average, while those with the more vigorous program improved 47 percent. So it works, whether you are by yourself or as a group, and more is better.

The researchers also tested different frequencies of exercise, asking whether five days a week is better than three. And the answer is no. Working out five days a week was not better than three. So, a longer total duration of exercise each week seems to be better than more modest exercise, but three days a week is enough.

It is important to note that not everyone improved. Overall, 46 percent of the fast exercisers had a significant improvement. For many of these, the depression effectively went away. But that also means that more than half the participants did not benefit. Even so, these results are similar to those of antidepressant medications or cognitive behavioral therapy. Many people don't get anything out of them, either.

Overall, these studies show that for many people, exercise helps alleviate depression. It can also help prevent it. The effect is not enormous, but it is real. Studies also show that, overall, exercisers are about 17 percent less likely to lapse into depression, compared with their sedentary friends, and that is true regardless of age; it works for young and old alike.[34]

If your joints or heart or general health are not up for a big walk, start small. A ten-minute brisk walk three times a week is fine. Then next week, make it fifteen minutes. Add five more minutes each week. Researchers at the University of Illinois found that a simple program of forty-minute brisk walks three times per week improved memory and reversed brain shrinkage.[35]

While you are at it, it pays to exercise outdoors. That's because sunlight is an antidepressant. When you get out into the sunshine, you can just feel it. For people who are depressed during the winter months—a condition called *seasonal affective disorder*—psychiatrists sometimes prescribe light therapy from specially designed boxes that provide a copy of sunlight. They work. If you can get the real thing—sunlight itself—by all means do.

Get Some Sleep

As great as exercise is, it is also important to rest. Sleep is good for your brain. A good night's sleep puts your mood on an even keel and also helps your memory. If you are not sleeping well, you will feel like your emotional control and memory are not what they should be.

To get a good night's rest, try these tips:

1. **Be careful about caffeine and alcohol.** Caffeine lingers a surprisingly long time in your bloodstream. For many people, traces of the caffeine from a morning cup of coffee are still circulating at bedtime. You might be able to doze off, but your sleep may be lighter and more easily disrupted. It's the same with alcohol. It can lull you into sleep, only to wake you up at 4:00 a.m. The reason is that, while you sleep, alcohol molecules transform to *aldehydes*, which are mild stimulants that can make your sleep rocky. At the risk of sounding like a party pooper, I have to say that there are advantages to skipping caffeine and alcohol, and better sleep is one of them.

2. **Get regular exercise.** If your muscles have not had much to do all day, you may find it hard to doze off and your sleep will be lighter. After all, part of the function of sleep is to allow the body to repair from the stresses and strains of the day. So, if you neglected your muscles all day, it is as if your body doesn't feel a need for much sleep, even if

your brain could really use it. If, at bedtime, you realize you have not had any exercise all day, try a few push-ups or squats. I am not suggesting that you play a set of tennis or run a marathon just before bed. The idea is just to tire your muscles so that your body will *demand* sleep.

3. **Stretch and yawn.** As children approach bedtime, they go through a ritual of stretching out their arms and giving a big yawn. Cats, dogs, and other animals do the same. I do not know exactly why this peculiar pre-sleep ritual primes the brain for sleep. But I have noticed that stressed out adults often do not do this, hoping that lying down and closing their eyes will do the job.

   Let me encourage you to try this simple exercise. As silly as it may sound, have a big stretch and a yawn about thirty minutes before you go to bed. The stretch and yawn can be entirely artificial—you're just going through the motions. But you will discover that a pretend yawn becomes a real one. If you do this about four times before you turn out the light, you will sleep better.

4. **Avoid prescription sleep medications.** While there are times when sleeping medications may be useful, I would encourage you to consider them a last resort. Zolpidem, the sleeping pill marked as *Ambien*, causes peculiar memory symptoms. Under its influence, people have done all manner of things, including walking and even driving, with no memory of the events the next day. Hopefully, with a better sleep routine, you will not feel a need for medications.

5. **If you awaken in the middle of the night and cannot get back to sleep, have a slice or two of bread or some other starchy food.** Starch appears to trigger the release of serotonin in the brain, allowing you to fall back to sleep.

April

April is fifty-four years old, living in Fort Defiance, Arizona, where she is the Director of Nutritive Services at a large medical center. I met April at a diabetes conference, where she described what had happened to her over the preceding year.

Her job is a busy one, she told me, but not busy enough to explain the gloomy mood she often found herself in. She was short-tempered with her staff, often blowing up about little things. "I was angry a lot, and I let my staff know," she said. "I didn't do it to be mean; I felt they weren't doing their jobs." She chided her boss about things she was neglecting.

Her daughter drew a parallel from the movie *The Devil Wears Prada*. "That's you, Mom! Don't be so mean. Your staffers are scared of you!"

"People didn't want to talk with me," April said. "It got to the point where I just stayed in my office and avoided everyone." It was the same with her family. Conversations drifted into arguments, and they were no longer enjoying their time together.

What April's staff members and family did not know was that she was not well *physically*. She was in pain. Her muscles and joints hurt constantly. Getting in and out of cars was painful. If she tried to go for a run or even a walk, she would end up drained and sore. When she got home from work, she just wanted to lie down. Her physical pain left her in terrible spirits.

She saw a doctor who examined her, got X-rays, and ran some blood tests that showed that her CRP level was elevated. CRP is a measurement of inflammation, as we saw earlier. It looked like she had rheumatoid arthritis.

She also had Meniere's disease—a disease of the inner ear that causes vertigo and hearing loss. She tried to reduce stress to diminish the symptoms.

And she needed to lose weight. Running a food service, she knew a lot about food and tried to focus her own diet on the guidelines of the USDA Pyramid. But it did not affect the number on the scale. Things were not getting better.

In October 2017, she learned something new. A low-fat vegan diet was being taught in her community to tackle diabetes, and many people were having great results, losing weight and getting their blood sugar under control.

Despite the fact that the winter holidays were approaching, she decided it was time to leave behind some not-so-healthful favorites. "I used to like nacho cheese with a lot of chili, or grilled cheese or cheese with fruits. It was hard to break these habits, but I did. I started to think about what I was going to eat and how it was going to make me feel."

She eased into the diet, and by January had crossed animal products and fried foods off the menu. It turned out to be easy to switch from meat to beans. Pinto beans and black beans with green chiles are now staples. She orders vegan pizza, and the local Chinese restaurant serves a delicious vegetable medley.

Within two weeks, she started to feel better. The pain in her muscles and joints began to ease. She was able to walk again without pain. Her CRP returned to normal. Her appetite became easier to control, and, bit by bit, she lost forty pounds. Heartburn, which had troubled her for years, went away. She sleeps more deeply now.

One day, a staffer surprised her with a question: "Why are you so happy all the time?" She was taken aback at first. But she realized that she felt better, and apparently it showed. "Because I don't have pain anymore," she replied.

"My staff has noticed the change," she said. "Because of pain, I was irritable. People used to get out of my way and try to avoid me. Now things don't bother me like they used to. I don't scold; I just ask. And they don't run away from me now."

Her family noticed the change, too. "We don't argue anymore. I enjoy being around them now." April's family has also started to follow her lead in the kitchen. Her sister makes potato salad with vegan mayonnaise and pasta salad without cheese. Her eighty-one-year-old mother started the diet, aiming to improve her diabetes. In the process, her A1C—a test that shows blood sugar control—dropped from an unhealthy 8.1 percent to 6.1 percent, and her walking has noticeably improved. Her father has started to make changes, too, bringing vegetables and fruits into his routine in a big way. April is delighted, because he had a triple bypass fourteen years ago, and a menu change will help keep him healthy.

"Now I think of other people who feel the way I felt," April said. For many, chronic pain and physical limitations affect their moods and interactions with others. As better foods restore good health, these problems start to melt away.

Foods for Better Moods

In addition to whatever treatment your doctor has prescribed, try the following simple steps for a healthier mood. By now, these steps will be familiar to you, as we have already seen how they apply to the mood issues of PMS and to other conditions:

1. Avoid animal products. You will want to avoid meats, dairy products, and eggs.
2. Favor high-fiber foods. That means beans, vegetables, fruits, and whole grains.
3. Keep oils low. It pays to use non-oil cooking methods, like steaming and baking, and to use oil-free dressings on salads.
4. Have plant-protein-rich foods, like tofu, tempeh, or beans, especially at breakfast.
5. Avoid sugar and chocolate. For many people, these foods seem to lead to irritability and depressed mood.

6. Minimize caffeine and alcohol. Caffeine and alcohol can be temporary mood-boosters. But what goes up, must come down, often leaving you feeling worse than before.

7. Take vitamin B_{12} daily. Your body needs only 2.4 mcg per day. So pick up the smallest supplement you can find at the drugstore or health food store, and you're set.

8. Exercise. Start slowly, if you have been sedentary, and work up to a forty-minute walk three times a week. Do more if you can. The Physical Activity Guidelines for Americans call for two and a half to five hours of moderate-intensity physical activity every week. If you prefer, substitute 75 to 150 minutes of vigorous activity each week.

9. Get plenty of sleep. When the clock strikes ten (or if you work evening or night shift, whatever is an equivalent time for you), turn out the light.

10. Sunlight. Sunlight boosts your mood. As it touches your skin, it produces vitamin D. More about that in the next chapter.

Feeling Good Again

Depression may be a response to psychological factors, and psychiatrists rely on talking therapies and antidepressant prescriptions. They have their role. But we now know that our moods are affected by physical factors, too, including food choices. A focus on plant-based foods, along with the other steps outlined in this chapter, has helped many people. Let me encourage you to see what it will do for you.

Part III

FEELING BETTER AGAIN

CHAPTER 12

A Healthy Diet

It is time to put the power of healthful foods to work. In this book, we have covered many different conditions. Some are caused by sex hormones getting off-kilter. In other cases, the issues relate to insulin or thyroid hormones. Happily, you do not need one diet to get estrogens into better balance and another diet to boost insulin sensitivity or to normalize your thyroid. Certain principles apply across the board, with just a few tweaks here and there.

In this chapter, we will look at foods to emphasize, foods to avoid, and how to make the process easy and fun. Let me encourage you to follow these steps carefully. For many people, this experience is life-changing.

While different people respond to foods slightly differently—for example, some people are more sensitive to the effects of oily foods and have to avoid them strictly, while others find they tolerate them without problems—these principles apply for more or less everyone. We will lay them out here and point out individual issues along the way.

Foods to Emphasize

As we have seen, plant-based diets get your hormones into a healthier zone. Plants have no animal fat, of course. They have essentially no cholesterol and are naturally high in fiber. As we saw in Chapter 1, "Foods for Fertility," they help you shed unwanted weight, and their fiber traps unwanted hormones in your digestive tract and carries them out with the wastes.

As you plan your meals, there are four groups to focus on:

Vegetables. Vegetables not only have you covered when it comes to boosting fiber and avoiding unnecessary fats. They do far more than that. Green leafy vegetables bring you calcium without the hazards of dairy products and iron without the hazards of meat. Perhaps surprisingly, many vegetables are rich in protein. Broccoli does not want to brag, but it is about one-third protein, as a percentage of its calories.

But what vegetables can really boast about are their antioxidants and cancer-fighting micronutrients. They are rich in folate, a B vitamin that is key to your anticancer defenses and may play a role in mood regulation. *Cruciferous* vegetables, like broccoli, kale, collards, cauliflower, and Brussels sprouts, activate your *phase 2* enzymes to help you eliminate toxins, as we saw in Chapter 3, "Tackling Cancer for Women." Orange vegetables, like carrots and sweet potatoes, are rich in beta-carotene, a powerful antioxidant that neutralizes free radicals that contribute to aging, DNA damage, and cancer. And don't forget vegetables from the sea, which bring your thyroid the iodine it was hoping for.

Rather than relegating vegetables to the sidelines, bring them front and center on your plate, and have more than one at a meal—a green and an orange one, for example. Broccoli and sweet potatoes or Brussels sprouts and carrots are a perfect match.

Choose organic, when you can. If it is not available, conventional produce is okay and always *far* more healthful than animal products.

Turning Vegetables into Candy

If green vegetables are not yet your thing, here's a tip. Steam up broccoli, kale, or Brussels sprouts, and cook them until they are soft (raw is fine for lettuce and cucumbers, but not for *cruciferous* vegetables, which should be well cooked). Then, top them with Bragg Liquid Aminos All Purpose Seasoning, a savory topping you'll find next to soy sauce at the health food store. Or try seasoned rice vinegar (an Asian seasoning) or lemon juice. These toppings soften the natural bitterness of the vegetables and make them delectable.

Fruit. Like vegetables, fruits are high in fiber, low in fat, and vitamin-rich. And, yes, fruits are sweet, but they are surprisingly gentle on your blood sugar.

Apples, bananas, oranges, pears, peaches, and grapes are familiar, of course, but try the full range: berries of all kinds, papayas, mangoes, cantaloupes, melons, watermelons, clementines, cherries, and whatever else has not been in your shopping cart for a while.

What the Colors Tell You

You have a built-in antioxidant-detection system that allows you to spot antioxidants from all the way across the room. It is your color vision. You see *lycopene* as the bright red color in tomatoes, watermelons, and pink grapefruit. Lycopene is a cousin of *beta-carotene*, the orange antioxidant in carrots and sweet potatoes. *Anthocyanins* are antioxidants that provide the deep colors in blueberries and grapes. The color in beets comes from an antioxidant of their own, called *betanin.*

(continued)

Note also that everyone finds these colors attractive (no one ever objected to the color of a tomato or a beet). Presumably, this system evolved to allow us to detect antioxidant-rich foods and to be attracted to them, and people who ate antioxidant-rich foods were more likely to survive and to pass along their DNA to future generations. Cats and other carnivores have a very different retinal structure—low on color vision and high on night vision and peripheral vision—perfect for hunting prey.

So while a cat's vision is great for spotting a mouse scurrying by at night, our color vision guides us to the colorful nutrients our bodies need. Of course, those same colors now end up in M&Ms bags, but they would prefer to guide you to antioxidants.

Let me encourage you to stock up with more fruit than you need and keep it on hand at home and at work. It is great for sharing, and when hunger strikes, a banana is a *much* healthier snack than string cheese, beef jerky, a cookie, or potato chips. Fruit is essentially fat-free, low in calories, and loaded with fiber and vitamins.

If you buy a cantaloupe or melon, cut it up and keep it in a bowl in the refrigerator. It will be ready when you are. And for a delicious dessert, mix blueberries, bananas, and chunks of papaya or mango.

Legumes. Legumes are beans, peas, and lentils—foods that grow in pods. They are rich in protein and calcium and lend themselves to endless traditional dishes from around the world: navy beans for chili, pintos for refried beans, chickpeas for hummus, lentils and split peas for soups, and soybeans (e.g., edamame, tofu, soymilk). Unlike animal products, which raise cholesterol and blood pressure, beans and other legumes have the opposite effects.

Beans are the undisputed fiber champions. As you will recall, fiber helps you eliminate excess hormones. That fiber also makes them filling and satisfying. If beans cause gassiness for you, start with small portions and cook them thoroughly.

Easier Than I Thought!

For the longest time, I was allergic to the idea of cooking beans from scratch. Who has the time? It was easier to just open a can. But then I discovered that beans did not require supervision. They essentially *cook themselves* and did not need much involvement from me in the process.

It's easy! Just put dried beans in a pot, rinse them, cover with water overnight, tossing in some salt, which softens the skins, and leave them alone. The next day, change the water and cook them until soft—about an hour. While they are busy cooking, you can do something else.

Once they are done, drain them and, if you like, blend them with a hand immersion blender, then top with salsa, jalapenos, or whatever you would like, to make a delicious dip or burrito stuffing.

Grains. Grains are staples for most traditional societies: rice in Asia; corn in Latin America and the American Southwest; wheat, oats, and millet in Europe; teff in Ethiopia and Eritrea; quinoa in Peru; and many others. They provide healthful complex carbohydrate for energy, along with protein and fiber. Like all plant foods, they are free of animal fat and cholesterol.

In their natural state, grains have an outer bran coating, which is removed when grains are refined (e.g., to make white bread or white rice). It is better to leave that bran coating intact, because that is where the hormone-taming fiber is. So whole-grain bread and brown rice are better choices. Two caveats: If white rice is all that is available at your favorite Chinese restaurant, go ahead and have it—it beats the socks off cheese or chicken. And if your whole-grain pasta turns out a bit too mushy (because fiber tends to hold water), white pasta is fine. It can be topped with chunky vegetables to replace the fiber that refining took away.

Creating Your Masterpiece

Vegetables, fruits, legumes, and grains are your palette; let's use these ingredients to create a meal masterpiece. If you are going Mediterranean, you might have lentil soup, followed by penne arrabbiata, along with grilled asparagus and fresh berries for dessert.*

A Midwestern meal might start with a spinach salad with slivered walnuts, then bean chili with kale and sweet potatoes, followed by applesauce for dessert.

A Mexican-style dinner could be a burrito of beans, rice, and veggies, topped with a salsa made from chopped tomatoes, onions, cilantro, and lemon juice, and served with baked plantains and sliced papaya.

There are endless possibilities. Page through the recipe section of this book, and see what calls to you. How about starting with Savory Dill & Potato Soup or Cauliflower Buffalo Chowder? You can keep it light with a Summer Panzanella Salad or a Thai Crunch Salad, and there are plenty of hearty dishes: Mediterranean Croquettes, Kung Pao Lettuce Wraps, Carrot & Pea Curry, Butternut Pasta, and many more. You'll find wonderful desserts, too. How about Chocolate Cupcakes or Apple Pie Nachos?

But I Don't Like to Cook

A plant-based diet will change your health, but it will not change your personality. So if you feel that life is too short to spend time in the kitchen, you'll still feel that way even if your kitchen is filled

---

* *Arrabbiata* sauce sounds like it is made from an Arab recipe. But it is actually the Italian word for "angry," calling to mind the spices used to make it. What you put it on might be easier to translate. *Penne* means "pens," as in a quill pen. *Capellini* means "little hair" or, better, "thin hair." *Linguine* means "little tongues," and *spaghetti* means "little strings." *Vermicelli*—well, let's not discuss that.

with healthful foods. Relax. Plant-based eating does not mean that you have to become a gourmet chef. If you are living on convenience foods, it is just a question of making the best choices. It takes no more time than you spend preparing food now.

Instead of the not-so-healthy frozen foods you might be familiar with, check out the huge array of vegan frozen foods—pizzas, enchiladas, curries, soups, precut vegetables, and everything else. You'll find ready-to-heat soups—butternut squash, minestrone, tomato bisque, split pea, and many more, often packed in BPA-free cans. The healthful vegan versions cook up as quickly as the unhealthy versions. Nearly every fast-food place has vegan meals, too, from veggie subs to meatless burgers, bean burritos, baked potatoes, and many others. At a Japanese restaurant, you might have miso soup, a green salad with ginger dressing, and sushi made with cucumbers and sweet potatoes.

An Indian restaurant would serve dal soup, a savory curry of spinach and corn on basmati rice, and mango slices for dessert.

That said, once you see what foods can do for you, you may find yourself attracted to the idea of preparing meals for yourself. Let Lindsay's recipes in this book tempt you.

Foods to Avoid

Dairy products. Milk was "designed" by nature to nourish a baby—cow's milk for calves and human milk for human babies. All animals are weaned, because milk's nutritional content is inappropriate at later ages.

Dairy-consuming adults subject themselves to all manner of things that are safe for the short period of breastfeeding but not for continued exposure. Dairy products are the leading source of saturated fat, which increases the risk of heart problems and Alzheimer's disease, and the only real source of lactose, the sugar that breaks down to release galactose. As you will recall, galactose is implicated in fertility problems and ovarian cancer. As we have seen, milk contains estrogens coming from a cow, and these estrogens end up in the products made from it. Although they are only traces, evidence suggests that they may indeed influence human biology.

Milk proteins appear to be common triggers for all manner of health conditions, including type 1 diabetes, rheumatoid arthritis, migraines, and many others. As we saw in Chapter 4, "Tackling Cancer for Men," milk-drinking also boosts your IGF-1 level; IGF-1 is linked to breast and prostate cancer.

Dairy products also tend to be a repository of whatever pesticides, industrial chemicals, and other contaminants the cow has ingested. At the same time, dairy products have none of the fiber you need to control hormones.

If it is calcium you are looking for, we will explore healthier sources below. And if you want something to splash on your cereal, you'll find soymilk, rice milk, almond milk, oat milk, and many other healthful varieties in the dairy case.

Nutritional Yeast

Nutritional yeast is a handy topping to get to know. It adds a mild cheesy flavor to pizza and works great in sauces, vegetable dishes, scrambled tofu, and soups. Some people sprinkle it on popcorn or mix it with ground cashews to make a pesto.

Although nutritional yeast tastes like cheese, nutritionally it could not be more different. Typical cheeses are 70 percent fat, as a percentage of calories. They are higher in sodium than potato chips and have as much cholesterol as a steak, not to mention estrogen traces. Nutritional yeast is fat-free, cholesterol-free, and estrogen-free; low in sodium; high in protein; and sometimes B_{12} fortified.

You will find it at health food stores in the supplement aisle or bulk aisle. Note, this is *nutritional* yeast, not baker's yeast or brewer's yeast.

Skip the Fat

Vegetable oils are healthier than animal fats. They are free of cholesterol and are usually low in saturated fat. Even so, it pays to keep them to a minimum, too. Here is why: Fats and oils have *9 calories in every gram.* That is true for animal fats, like butter and chicken fat, and for vegetable oils, too. For comparison, sugar has only 4 calories per gram. So gram for gram, fats and oils have more than twice the calories in sugar. In fact, that's the case for all carbohydrates, whether they are found in bread, potatoes, fruits, candy, or any other starchy or sweet food. They have only 4 calories per gram.

So if weight loss or controlling hormones is your goal, you will find that limiting oils really helps. It makes a difference for other conditions, too. Robin, whom we met in Chapter 2, "Curing Cramps and Premenstrual Syndrome," found that avoiding oily foods was a key step in conquering her menstrual pain. Similarly, Nina and Randa, whom we met in Chapter 10, "Healthy Skin and Hair," found that their acne was triggered by oily foods that most people would consider to be normal parts of the diet. Avoiding these foods kept their skin healthy.

If you have eliminated animal products and boosted high-fiber vegetables, fruits, and beans, and have still not met your health goals, I would suggest being very cautious about oils. That means avoiding added oils and fatty foods (e.g., nuts, nut butters, and guacamole), and reading package labels.

What About Olive Oil?

You may have imagined olive oil is a health food. It has certainly been marketed that way. Perhaps you were seduced by the clever marketer who came up with the term *extra virgin*. It sounds so much sexier than "cold mechanical extraction."

(continued)

But there is no faucet on an olive tree. To get olive oil, producers take thousands of olives, throw away the fiber and pulp, and concentrate the oil. Just as sugar is extracted from sugar cane, olive oil is a concentrated extract. It is something Mother Nature never imagined anyone would eat, except in tiny traces in olives.

It should be said that, nutritionally, olive oil beats animal fats any day. It is only 14 percent saturated fat. Compare that with chicken fat (30 percent), lard (39 percent), or butter (63 percent). But 14 percent is not as good as zero, which is what you get if you leave the cap on the bottle. And, with 9 calories in every gram, dipping your bread into olive oil has the same effect on your waistline as dipping it into chicken fat, lard, butter, or Vaseline, for that matter.

Here are some tips to help you cut the fat:

1. Avoiding animal products allows you to avoid all animal fat, of course.
2. In the plant kingdom, the main fatty foods are added oils, nuts, seeds, and avocados. If you are looking to lose weight or tackle a hormonal issue, it pays to avoid them.
3. For packaged products, check the label and favor those with no more than about 3 grams of fat per serving. This is very low and a good place to start until you know whether you can tolerate a more generous fat content.
4. Instead of sautéing onions or garlic in oil, cook them in water, vegetable broth, wine, or even a dry pan.
5. Steamed, boiled, or baked foods are lower in fat than fried foods.
6. On toast or potatoes, skip margarine and butter. Have nonfat toppings instead: jam, jelly, or cinnamon on toast, and pepper, mustard, salsa, soy sauce, or chopped veggies on potatoes.

7. Instead of oily salad dressings, try lemon juice, flavorful vinegars (balsamic, apple cider, or seasoned rice vinegars), or nonfat dressings (you will see many on grocery shelves).

Skip Coconut and Palm Oils

When partially hydrogenated (trans) fats were shown to raise cholesterol levels and were banished from many foods, manufacturers looked for a replacement with a similar buttery mouthfeel and a long shelf life. And coconut and palm oils fit the bill nicely.

The problem is, despite their rosy claims, they are extremely high in saturated fat and will raise your blood cholesterol, too. They are fine for waxing your shoes, but do not eat them. Read labels and avoid foods with coconut or palm oil. They often turn up in peanut butter, cream cheese substitutes, and other thick, fatty products. You'll also see them in some veggie burgers that claim to have an impossibly meaty taste. But there are far better choices, including plenty of healthful veggie burgers with no saturated fat at all.

You have probably heard of the controversy over palm oil production, which has led to massive deforestation in Indonesia, devastating the orangutan population. There is no good reason to consume these products.

Transition Foods

As you move away from meat and dairy products, you can take comfort in products that provide the tastes and textures of these foods, without the regrets. You will find substitutes for bacon, sausage, burgers, and hot dogs made from soy, wheat gluten, or other plant-based ingredients. Try different brands and see which ones you like best.

You will also find cheeses made of cashews or almond milk. As a group, vegan cheeses are far healthier than the animal-derived foods they replace. Even so, most are still high in fat. Their main value is as a transition food and as an occasional treat for party guests.

Complete Nutrition

A plant-based diet will bring you far better nutrition, compared with a meat-based diet. Vegetables, fruits, legumes, and whole grains bring you fiber and other nutrients you won't find in animal products. They also have protein, healthful complex carbohydrate, and traces of the healthful fats you need. Even so, there are some common questions that we'll want to address:

Protein. Your body uses protein to repair daily wear and tear and to build various molecules your body uses throughout the day. So where do we find it? Many of us grew up with the idea that meat is the protein source on our plate, while vegetables provide vitamins and rice or potatoes give us carbohydrate. However, protein is actually in many foods. Let me give you some numbers:

According to the U.S. government, a woman should aim for about 46 grams of protein per day, and a man should aim for about 56 grams.

What would happen if, on a given day, just as an experiment, you were to have nothing but broccoli all day? Let's say you eat 2,000 calories a day, which is a typical amount. If all those calories came from broccoli, it turns out that you would get *146 grams of pure protein*. The next day, let's say you were to have nothing but lentils all day; 2,000 calories' worth would deliver 157 grams of protein. Obviously, no one would eat just one food in a day. But what if you had some broccoli and some lentils, along with various other foods from plant sources? There is protein in the broad array of vegetables, legumes, grains, and fruits. Nutrient analyses show that plants easily provide the amount of protein the government calls for.

Surprising, isn't it? There is a lot of protein in plants, which is why cows, horses, elephants, and giraffes are able to build such enormous bodies. Even high-level athletes easily get the protein they need from any normal varied diet from plant sources.

Iron. Your red blood cells use iron to make hemoglobin, which transports oxygen to wherever your body needs it. Plants provide iron in its most useful form, called *non-heme iron*. With this form of iron, your body can absorb more when you are running low on iron and absorb less if you already have plenty of iron on board. That is good, because iron is harmful when overconsumed; it sparks the production of free radicals, molecules that can damage your arteries, your skin, and all the rest of you. Excess iron can also contribute to heart disease, cancer, premature aging, and Alzheimer's disease.

How would you ever get too much? Meat has a form of iron called *heme iron*, which defies your attempts to regulate its absorption. It just barges into your body whether you want it or not.

So it is better to get your iron from plants. Especially good sources are "greens and beans"—that is, green leafy vegetables and legumes. If you are low in iron, you can boost absorption with vitamin C–rich vegetables and fruits. Dairy products impair iron absorption.

Calcium. Calcium is used to build bones, of course, and it plays other roles in the body, too. Most people think of dairy products as *the* source of calcium. But it is important to understand that cows do not make calcium. Calcium is an element in the earth. It passes from the ground into the roots of grass and other plants, and a cow eating grass gets that calcium and passes some of it into her milk. If you were to drink milk, you would absorb about 32 percent of its calcium. The other 68 percent is not absorbed.

If instead you were to eat green plants directly—hopefully not grass, but broccoli, kale, collards, Brussels sprouts, or other greens—you would get calcium directly, with a higher absorption fraction, typically around 50 percent (although spinach is an exception, with low calcium absorption). Beans are a great source, too, just as is the case for iron. While milk's calcium comes along with unhealthful saturated fat and lactose sugar, green vegetables bring you calcium along with beta-carotene, iron, and healthful fiber. They are far healthier sources.

To absorb calcium, you need vitamin D (see below).

Vitamin B$_{12}$. You need vitamin B$_{12}$ for healthy nerves and healthy blood. Without it, you can become anemic and your nerves can malfunction. But B$_{12}$ is not made by animals or plants. It is made by bacteria. Some have speculated that prior to the advent of modern hygiene, traces of bacteria in soil, on plants, on our fingers, or in our mouths produced the traces of B$_{12}$ we needed. Whether that is true or not, it is certainly not the case today.

Meat-eaters get some vitamin B$_{12}$ in their diets, because the bacteria in a cow's gut make it and it passes into meat and milk. Our digestive tracts have bacteria that make B$_{12}$, too, although it appears that they are so far along in the digestive tract that the B$_{12}$ they make is not absorbed. But even meat-eaters can run low in vitamin B$_{12}$, because they may not absorb it adequately. That is especially true if they are over fifty, do not produce enough stomach acid, or take acid-suppressing medications or metformin, a common diabetes drug. As a result, health authorities recommend that everyone over age fifty take a B$_{12}$ supplement or use B$_{12}$-fortified foods. This is actually good advice for everyone, no matter what your age. And a B$_{12}$ supplement is essential for anyone on a vegan diet.

It is easy to find. It is in all common multiple vitamins and many fortified foods. The best choice is to pick up a bottle of vitamin B$_{12}$ at the drugstore or health food store. They all exceed the 2.4 mcg you need, and it is not toxic in higher amounts.

Two Other Supplements to Consider

There are two other supplements to think about:

Vitamin D. Vitamin D helps you absorb calcium, and it also helps prevent cancer. Normally, sunlight on your skin produces the vitamin D you need. Had we all stayed in equatorial Africa where our species was born, we would all get plenty of it. Just twenty minutes of sunlight on your face and arms each day gives you the vitamin D you need. However, those restless humans who wandered to places like Seattle and Reykjavik found that sunlight was less predictable. Without regular sun exposure, it is easy to run low in vitamin D. Sunscreen can also block vitamin D production. In these circumstances, a vitamin D supplement has you covered. Most experts recommend about 2,000 IU per day.

EPA and DHA. Eicosapentaenoic acid (EPA) and docosahexaenoic acid (DHA) are omega-3 fatty acids that have many roles in the body. We described them briefly in Chapter 4. In theory, your body makes them from ALA (alpha-linolenic acid), which is available in many plants, especially walnuts, almonds, green vegetables, and ground flaxseeds.

However, the enzymes that lengthen the ALA molecule into EPA and DHA are easily distracted by other fatty acids that come their way. So if you eat fatty foods, they can tie up these enzymes and slow your production of EPA and DHA. Fish-eaters get some DHA but also get a fair amount of saturated fat, cholesterol, and pollutants along with it. So some people choose to take EPA and DHA supplements.

Although health claims for these supplements have not held up very well,[1] the best argument in their favor is that running low in DHA in particular might increase the risk of Alzheimer's disease.[2] The argument against them is that men with higher omega-3 blood levels have been shown in some studies to have a higher risk of prostate cancer, as we saw in Chapter 4. We do not yet know if this will be confirmed in additional studies or if there is an analogous danger for women.

In case it helps you to decide whether or not to supplement with EPA and DHA, OmegaQuant makes an easy mail-in test kit that determines your current EPA and DHA levels from a drop of blood and helps you understand if you are in a healthy range or not. Just search by the company name online.

If you do decide to take an EPA/DHA supplement, I suggest choosing a vegan version, rather than fish oil. They are free of the concerns about impurities and odor raised by many fish oil brands, not to mention being better for the fish and the environment, and they are widely available online.

Choosing the Best Carbohydrates

The carbohydrate group includes starches and sugars. As they digest, they release sugar into the bloodstream.

It is fashionable these days to blame sugar for weight gain, diabetes, and all manner of other health problems. And it is certainly true that a sugary soda is not a health food, even it is *Dr* Pepper. But remember that sugar—that is, glucose—is actually the main fuel for your brain, muscles, and all the rest of you. As you know by now, sugar has only 4 calories per gram, compared with 9 for chicken fat, beef fat, fish oil, and cooking oils. It is not especially fattening. If you thought that sodas are responsible for the obesity epidemic, the real weight-gain culprits might be the chicken nuggets, cheeseburgers, and greasy fries served with them. Similarly, the butter or shortening baked into cookies and cakes is much more calorie-packed than whatever sugar they hold.

That said, some carbohydrates are better than others. Here is a quick reminder about foods with a high Glycemic Index—that is, foods that tend to raise your blood sugar—and some helpful replacements:

- Table sugar tends to raise blood sugar, while fruit is much gentler on your blood sugar.
- Wheat breads have a high GI. Rye and pumpernickel are better choices.
- Cold cereals tend to raise blood sugar much more than bran cereal or old-fashioned oatmeal.
- Instead of typical white potatoes, favor sweet potatoes.

In addition, beans are always healthful, low-GI foods, and pasta—even white pasta—is gentler on your blood sugar, compared with other flour products, simply because it is compact and slower to digest than bread.

Alcohol

As we saw in Chapter 3, alcohol raises the risk for breast cancer, and it does the same for cancer of the colon, rectum, pancreas, and other organs. While some research has suggested that occasional alcohol use may be associated with reduced risk of heart problems and Alzheimer's disease, it is not at all clear that this is cause-and-effect or that benefits would even be apparent in people who are following an otherwise healthy diet and lifestyle. The less you drink, the better.

What About Salt?

Salt—that is, sodium—is an essential nutrient for the body. But in excess, it raises blood pressure and also causes you to lose calcium in your urine. Most health authorities recommend limiting sodium to 1,500 to 2,000 mg per day, which works out to roughly 600 mg for an average meal.

In their natural state, vegetables, fruits, whole grains, and legumes are very low in sodium. Manufacturers often add salt to commercial products, catering to consumers' tastes. Low-sodium brands are widely available.

How to Start

As you will see, I have recommended chucking out meat, dairy products, and some other foods that may have been familiar staples for you up to now. But not to worry. Having worked with thousands of participants in our research studies, we have developed a way to smooth this transition. I have never seen anyone unable to do it. We will break it into two steps.

Step 1. Take a week and, during this time, explore the plant-based options available to you. Do not feel a need to take anything out of your diet for now. Rather, our goal at the moment is simply to get to know healthier choices.

On a piece of paper, jot down headings for breakfasts, lunches, dinners, and snacks, and, over the next seven days, under each heading, fill in foods that are free of animal products that would fit the bill. Your job this week is to try these foods and see which ones you like best.

Let's start with breakfast. If you have always had milk on your cereal, this might be the week to try almond milk, soymilk, or another variety of plant-based milks. Pick up some old-fashioned oatmeal and see which toppings work best for you: cinnamon, blueberries, strawberries, or whatever grabs you. If scrambled eggs have been your breakfast of choice, now could be the time to try scrambled tofu, which happens to have almost exactly the look and texture of egg white and takes up the flavors it is cooked with. And try the delicious breakfasts in this book: Perfect Pancakes, Breakfast Pilaf, Strawberry Shortcake Polenta, and many more.

For lunch, you might try various soups, from lentil and split pea to tomato bisque and miso soup. Or try vegetable chili, bean burritos, veggie burgers, veggie hot dogs—see what calls to you.

Same for dinner. Whether you eat at home or at restaurants, your job this week is to seek out vegan choices and see which ones you like best. In the recipe section, you will find Mongolian Vegetable Stir-Fry, Moroccan Pizzas, Roasted Quinoa Pie, Southwest Lentil Mac, Shanghai Noodles, and many other delicious meals.

For snacks, you will want to get to know fresh fruit again if you have been neglecting it. It is always a winner.

Simply jot down the foods that you like for each category.

Step 2. Now that you know your favorites, your job is to actually eat these foods for the next three weeks. During this time, avoid animal products completely so that you can follow a completely vegan diet over the next twenty-one days. If you do this well, two things will happen.

First, you will feel healthier physically. You will likely shed a few pounds. If you have high blood sugar, it will start to fall, as will cholesterol and blood pressure. If you have sore joints, they may start to feel better. It might take a bit more time for hormonal issues to resolve, but three weeks will give you a solid start on a new path.

Second, your tastes will change. You will find that you no longer miss meat and cheese as much as you might have thought, and your desire for fatty foods will diminish. This taste reboot will not work as well if you have animal products every few days; that will only rekindle your tastes for the foods that have gotten you into trouble. That's why we do it "all vegan, all the time" for three weeks.

Let me draw an example from my own childhood; you may have had this experience, too. One day, my mother announced that we would no longer have whole milk in the house. It was to be skim from then on. At first, skim milk tasted watery and was not palatable at all. It did not even look right. But soon we adjusted to the lower-fat taste. A few weeks later, if we happened to taste regular milk again, it seemed too thick and unpalatable.

The point here is not that skim milk is health food. It is not; you are much better off with plant-based milks or just a glass of water. But I use this example to show that tastes change rapidly, based on the foods that you are in contact with.

After three weeks on a healthy vegan diet, you will find the prospect of healthful eating much more approachable. If you stick with it, you will find the experience very rewarding and even life-changing.

If by chance, you do fall off the wagon, just dust yourself off and get right back on. Don't let a momentary indiscretion become a lasting problem.

Support and Motivation

It is always easier to change any habit when we have the support of those around us. So let me encourage you to invite family, friends, or coworkers to join you as you change your diet, perhaps just as a three-week experiment. You will be surprised at how many people are game for trying something relatively quick that might improve their health, and how a short experience like this can open their eyes.

If family members do not want to join you, they can still help. Let them know that their encouragement will mean a lot to you and ask them to help you steer clear of the things you are trying to avoid.

It also helps to take advantage of other motivators. If you have a partner, spouse, or children, your good health is important to them. Keeping that in mind can be a strong motivator. Many people are moved by how a plant-based diet helps the planet or the animals we share it with. These issues are important and often have more power to keep us on the straight and narrow than health concerns alone.

On the Road and on the Town

When you dine out, you will find lots of vegan choices. Here are some helpful tips:

- Think international. Italian, Mexican, Chinese, Japanese, Thai, Vietnamese, Korean, Indian, and Ethiopian restaurants all have many plant-based choices.
- Ask for oil-free foods. Although restaurateurs can easily prepare vegan meals, they have more trouble containing their exuberance with oil. Ask that your food be prepared without oil.
- If you are at a diner for breakfast, ask your server to have the cook throw some mushrooms, tomato slices, onions, spinach, or asparagus on the grill, either separately or mixed together. No one will bat an eyelash, because they cook these ingredients for omelets. You are just skipping the eggs. Add rye toast (no butter), oatmeal, or grits on the side. Many diners also serve veggie burgers topped with mushrooms, onions, tomato, and lettuce any time of day.
- If it's fast food you are looking for, submarine sandwich shops can easily build a sandwich from lettuce, tomato, spinach, cucumbers, olives, pickles, and hot peppers, drizzled with red wine vinegar, on a toasted bun. As I mentioned earlier, taco restaurants offer bean burritos, and many burger outlets have veggie burgers or baked potatoes.
- When you travel, many hotels, especially midrange hotels, keep a microwave and refrigerator in each room. Extra points if you rent a kitchenette. You can stock up at a nearby grocery store, and you will be set.

Troubleshooting

Sometimes things don't go quite as you planned. Let's look at common ways things go off track, and how to fix them.

Not Losing Weight Fast Enough

On a low-fat, plant-based diet, weight loss is typically one half to one pound per week. Some people lose weight a bit faster, others a bit slower, but that is a good average. If that sounds modest, keep in mind that there are fifty-two weeks in a year, and if you stick with it, that lost weight will never come back. If your weight loss is somewhere in that range, you are doing fine. But if you are not losing weight at all or are losing at a trudging pace, take these steps:

1. Check your weight. First, let's see what your weight loss really is. Weigh yourself once per week on a reliable scale at about the same time of day. If you are losing *any weight at all* each week, you are doing fine. Keep in mind, this weight loss is a one-way street. If you do not lose weight one week, do not worry about it. Your weight will bounce around a bit. However, if you do not lose weight for two weeks in a row, it's time to take action. Do not rationalize this by saying you are losing inches or you feel better. If you are overweight and the weight is not coming off, that means the diet is not working. It is time to go on to the next step.
2. Recheck the basics. Are your meals free of animal products? If you answered, "Well, I'm about 90 percent there," you have found your problem. Surprising as it may sound, that other 10 percent is holding you back. Also, check the fat content of the foods you eat and toss out those with more

than 3 grams per serving. At restaurants, be sure to ask your server to minimize oils.

3. Keep it simple. Emphasize vegetables, fruits, whole grains, and legumes and de-emphasize processed foods. That way, your meals will retain their natural fiber and will be less likely to have hidden fats.

4. Have more raw foods. Raw foods tend to accelerate weight loss. Have fruit as a snack, and have a salad of greens, cucumbers, and tomato slices as a meal-starter.

Losing Weight Too Fast

If you are losing weight so quickly that you are worried that you might eventually just disappear, take heart. Once you get closer to your ideal body weight, your weight loss will slow down. If you are already at your ideal body weight and your friends are commenting that you are getting too thin, check your BMI. If it is between 18.5 and 25 kg/m^2, you are fine. Your real problem is that your friends are comparing your weight to theirs. But if you are indeed dropping below the healthy weight range, have more food (e.g., fruits, vegetables, grains, and beans). Stick with healthy choices but increase your quantity.

If unwanted weight loss persists, see your doctor to see if there is an underlying reason that needs to be addressed.

Persistent Hunger

When we were children, we allowed ourselves to be hungry some of the time. A few hours after breakfast, hunger would arrive, and we would live with it for an hour or so before lunch. The same thing happened before dinner. That is normal; momentary hunger is nothing to fear. Many adults wipe out that cycle, eating whenever the slightest thought of hunger arrives. If that is you, be a bit slower to open the refrigerator.

On the other hand, if you are hungry much of the time, it is time to eat larger quantities at mealtime and to have healthy snacks. Keep plenty of fruit on hand.

If you are experiencing cravings, rather than hunger—in other words, you are drawn to food but are not actually hungry—this is the common result of being surrounded by highly palatable foods. Sugar and chocolate might call your name, and if you have small amounts of these foods once in a while, there is no reason to worry. But if you are having them every day (or several times a day), they can slow your progress.

Sometimes unhealthy snacks call your name on a certain schedule. You have a bit of chocolate now, and it is as if your brain then sets a timer to remind you to have chocolate again twenty-four hours from now. Whatever unhealthy food you eat now primes your appetite to want that same food tomorrow. By breaking out of this cycle, you can turn that timer off. Again, be sure to have healthy snacks on hand so that hunger does not drive you to less-than-healthy choices.

Sharmila

"I started this diet only because I had to." True enough, Sharmila was forced into a menu rethink. And she is thrilled.

Sharmila grew up in Bangalore, India, and today she lives in Chicago with her husband, two daughters, and her parents and mother-in-law.

"One day, my teenage daughter announced that she was going vegan," Sharmila said. "She had learned about the cruelties in dairy farming and wanted nothing more to do with it." A few months later, her husband made more or less the same diet change. He had developed an unusual autoimmune condition that caused a sudden loss of vision. Fortunately, the condition, called Vogt-Koyanagi-Harada disease, was quickly diagnosed. With massive doses of steroids, he recovered his eyesight. But in the wake of that medical scare, he started looking into the causes of autoimmune diseases and found that foods, particularly dairy products, have been implicated as triggers. So, like his daughter, he adopted a vegan diet, and his health rebounded.

Thus Sharmila ended up with two family members who had gone vegan. From a health standpoint, the diet change made complete sense. But that did not mean the kitchen transition was easy. She found herself cooking different meals for different family members, which was a chore, to say the least. "I was used to cooking with ghee (clarified butter) and yogurt and found it hard to re-create recipes," she said. After a fair amount of experimenting, she decided that the whole family should share the same healthy diet, and she would make vegan foods for the whole family. At first, she and the family dropped animal products but did not avoid oily or processed foods. That was a start.

At the time, Sharmila had health issues of her own. She had had an irregular menstrual cycle all her life, with menstrual pain that had been increasingly unbearable. Her doctor performed an ultrasound, finding bilateral ovarian cysts and uterine fibroids.

Sharmila and her family took additional steps. They avoided animal products, and they knocked out the oily and processed items. After six months on this healthy regimen, the radiologist repeated the ultrasound and was stunned by the change. The cysts were gone, and fibroids had not grown any further. He reported the change to her gynecologist, who spent the next twenty minutes asking what she had done to make her cysts disappear. Sharmila told her about her diet change and how her pain and clots had reduced, too.

Today, she has boundless energy. She enjoys running and can go for eight miles at a stretch. "I love cooking and am constantly re-creating whole-food, plant-based, no-oil foods," she said. "My best friend is my air fryer. I use soy yogurt, too. I can re-create any dish you want!"

There is always some awkwardness in learning new things. But soon, a healthy diet becomes second nature, and the payoff is enormous. See what it will do for you.

CHAPTER 13

Avoiding Environmental Chemicals

Progresso soup. It really seems like a notch above the other brands, doesn't it? It sounds Italian, the cans are big, and you don't need to add water. But Progresso soups were part of an experiment with some surprising results.

Researchers at the Harvard School of Public Health asked volunteers to have a canned Progresso vegetable soup every day for five days and tested their urine before and after. At the end of the experiment, the researchers found a chemical called *bisphenol A* (BPA) at more than ten times the usual level. When researchers did the same experiment with freshly made soup (as opposed to the canned soup), BPA levels did not increase.[1] This finding was troubling, because BPA has been linked with diabetes, heart disease, and liver disorders.[2] So how did it get in the soup?

BPA is used in the resins that line food cans—including the soup cans used by Progresso and other major labels. It is also used in hard, clear plastics, particularly reusable plastic bottles. Manufacturers love it, because it is tough and heat-resistant.

Environmental health scientists are not so keen on it, however. From cans and bottles, it leaches into the foods or beverages we consume, and it is rapidly absorbed into our bodies.

If that sounds creepy, here's something even more troubling: Let's say you go to a health food store and buy soup in a can that is specifically marked "BPA-free." You walk out the door with your credit card slip in hand and, unbeknownst to you, the thermal paper used for the credit card slip was coated with BPA, which is passing through the skin of your fingers and into your bloodstream before you get back to your car.

BPA is in the receipts printed out at cash registers, gas pumps, and ATMs. It is in boarding passes, lottery tickets, the receipt your laundry attendant gives you when you drop off your dry cleaning, and more or less any other piece of printed paper that comes out of a machine. You can feel the residue on your fingers. If you scratch the printed side of the paper with your thumbnail, a dark streak shows you it is thermal paper.

So the Harvard research team did a new study. No Progresso soup this time. Instead, they asked volunteers to print and handle cash register receipts over a two-hour period. If they wore gloves, nothing happened. There was no absorption of BPA. But when the volunteers handled receipts without gloves, the BPA passed straight through their skin and into their bloodstreams. Urine samples showed that BPA levels increased fivefold over the next four to twelve hours.[3] So, yes, that credit card slip in the palm of your hand passes BPA through your skin and into your bloodstream.

What does it do? Researchers at Kaiser Permanente in Oakland, California, zeroed in on the hormonal effects of BPA exposure in Chinese factory workers. They found that men who had BPA traces in their urine were much more likely to have lower sperm counts and worse sperm motility, compared to men with no BPA exposure.[4]

BPA-exposed workers also had a higher incidence of sexual dysfunction (erectile dysfunction, loss of libido).[5] These effects were not life-threatening, of course. But they raised the possibility of more serious hormonal effects, including hormone-related cancers. In later studies, BPA exposure was indeed associated with hormonal changes in men—a combination of higher testosterone and higher estradiol blood levels.[6] Among women, researchers have linked BPA exposure to higher estradiol levels[7] and a higher risk of polycystic ovary syndrome.[8]

In a 2012 *JAMA* study, researchers from New York University checked BPA levels in urine samples from 2,838 children, aged six through nineteen. Results showed that the more BPA there was in their urine, the more likely the children were to be obese. Those with the highest BPA levels had more than double the obesity risk, compared with those with the lowest BPA exposure.[9]

In all of these areas, however, study findings are mixed. Some have confirmed problems with BPA, while others have not, leaving it an open question as to what BPA's real effects may be. The U.S. Food and Drug Administration banned the use of BPA in baby bottles and sippy cups, recognizing that BPA can enter the cells of the body and attach to estrogen receptors and that that could be bad for a developing baby. But governmental authorities have held that risks to adults, if any, are not serious.

My advice—regardless of your age—is to play it safe. Apparently, Progresso is taking a similar attitude. Its website says, "While there is significant agreement among scientific and government bodies worldwide that BPA-lined cans do not pose a risk to your health, we know that some consumers prefer non-BPA lined cans. That is why, since 2016, we have shipped more than 25 million non-BPA lined cans and will continue to transition into non-BPA lined cans." You, the consumer, do not need to wait. Below, I'll show you how to reduce your exposure.

Better Living Through Chemistry?

We have certainly benefited from the chemicals used for medications and many consumer products. On the other hand, we don't always want extra chemicals in our lives—in our breakfast or lunch, for example. The fact is, hundreds of potentially toxic chemicals end up in foods. Sometimes, their dangers have been obvious. Other times, their effects are harder to spot.

BPA is just the tip of the chemical iceberg. Let us take a look at other common chemical exposures and how to keep yourself and your family safe. My goal is not to frighten you or to make you think you need a PhD in chemistry in order to do your shopping. As you will see, there are some common themes and easy steps that you can take right now.

Phthalates

Phthalates (the *ph* is silent) are used to make plastics flexible, so they are less likely to crack and break. Think, for example, of how a plastic raincoat or shower curtain bends, while a plastic pen does not. Phthalates have a zillion other uses, too, and end up in inflatable toys, garden hoses, vinyl flooring, automotive parts, adhesives, and personal care products like soaps, shampoos, and nail polish. They are used in paint, modeling clay, and in the coatings of some pills.

Here is the issue: Phthalates are not firmly bound to the molecules that make up plastic and can easily leach out, particularly when heated. Picture a plastic water bottle left in a hot car. The phthalates pass easily into bottled water.[10] You cannot detect it by taste or smell. They are so widespread that, if you were to check urine samples, you would find phthalate traces in most people.

The biggest source is not bottled water, however. The biggest source is food. Phthalates in containers dissolve especially well into anything fatty, like milk, butter, meat, and cheese.[11] And they abound in fast foods. A 2016 study from the George Washington University found that the more often adolescents or adults ate fast food, the higher the phthalate levels in their bodies.[12] Of course, even without phthalates, there are enough hormonal effects from cheese, chicken, and beef to knock your hormones out of whack, as we have seen in previous chapters.

In children, phthalates have been linked to developmental and behavioral problems,[13,14] insulin resistance,[15] higher blood pressure,[16] and allergies. In adults, phthalates have been linked to diabetes[17] and metabolic syndrome—a combination of problems with body weight, blood pressure, blood sugar, and blood lipids.[18] At the end of this chapter, we will look at ways to minimize our risk.

Pesticides

Pesticides are used on lawns, in parks, and especially on agricultural lands. When it comes to killing pests, they really work. And, of course, what kills a "pest" may also have health effects for you. In World War II, *dichlorodiphenyltrichloroethane* (DDT) was used to control typhus, malaria, and dengue fever. When the war ended, it became commercially available for killing insects on farms and eradicating malaria-carrying mosquitoes. However, DDT kills beneficial insects, too, as well as birds and fish. And it bioaccumulates—that is, it builds up in the body, particularly in body fat—so small doses over time can be increasingly toxic.

DDT is an *endocrine disruptor*—a chemical that wreaks havoc with your natural hormones. Soon after it was released to the market, studies raised concerns about its effects on menstrual function, fertility, pregnancy, infant growth, thyroid function, and cancer risk. In 1972, DDT was banned.

Although DDT is no longer in commercial use, many other pesticides still are. Monsanto's *Roundup*, for example, is the most popular weed killer in the United States. Its active ingredient, *glyphosate*, kills plants by blocking the enzymes they need for growth. It kills not just weeds, but other plants, too. So Monsanto sells genetically engineered "Roundup Ready" crops that can survive Roundup. Monsanto's idea is that farmers plant Roundup Ready seeds and then use Roundup to kill other plants that try to invade.

Could it hurt you? Maybe. Test-tube studies have shown that when mixed with breast cancer cells, glyphosate attaches to their estrogen receptors and pushes the cells to proliferate. In other words, it acts like an estrogen.[19] In 2015, the International Agency for Research on Cancer of the World Health Organization determined that glyphosate is "probably carcinogenic to humans," particularly noting the risk for non-Hodgkin's lymphoma, a common form of blood cancer.[20] The European Food Safety Authority disagreed, calling glyphosate safe, at least so far as cancer risk is concerned.[21]

Bottom line: Roundup's risks are still not entirely clear. So Monsanto may be Roundup-ready, but you and your children may not be. See below for ways to minimize your exposure.

The second-leading herbicide in America is *atrazine*. It is widely used as a weed killer for corn and sorghum crops used for animal feed. It is also sprayed on lawns and golf courses. As you can imagine, it gets on plants and soil—after all, that's the whole point of an herbicide. And because it trickles into rivers and streams, it can end up in the last place you would want it—in your drinking water. Researchers at the University of Illinois School of Public Health in Chicago tracked atrazine concentrations in the water in twenty-two Ohio communities and looked at the pregnancies that occurred in these communities, finding that atrazine in drinking water was associated with low birth weight. Their data also showed that atrazine levels soar in May and June, reaching levels more than seven times higher than those in early spring, reflecting the time of year when atrazine is applied.[22]

Bottled water is typically atrazine-free. And a reverse osmosis filter system—or even a typical Brita faucet filter—will remove atrazine, although other filters (e.g., pitcher filters) may not.

There are, of course, many other pesticides. Researchers from the University of London's Centre for Toxicology tested thirty-seven pesticides for their effects on human cells.[23] Of these, thirty had hormonal effects, including seven that were *androgenic*, meaning they behaved like male hormones, and twenty-three that were *antiandrogenic*, meaning they interfered with the effects of male hormones. In women, pesticide exposures have been linked to infertility, menstrual disturbances, stillbirths, and developmental defects in their children.[24]

And That's Just for Starters

There are many other chemical exposures that are part of our day-to-day lives. Polychlorinated biphenyls (PCBs) are synthetic chemicals whose production was banned in the United States in 1977. Decades later, they persist in the environment and end up in fatty fish and other animal products (dairy products, meat, and eggs) and are frequently detectable in human tissues.[25] PCBs can harm your immune, reproductive, nervous, and endocrine systems. They can also cross the placenta and contribute to cognitive problems in growing babies,[26] and they can enter breast milk.[27] They are not a reason to choose formula over breast milk, but they are a reason for women to follow as clean a diet as possible well before they even think about becoming pregnant.

Dioxins are by-products of many industrial processes.[28] Like PCBs, they end up in the body tissues of animals—including the human animal. Once ingested, they remain in your body for years and can affect your immune, reproductive, nervous, and endocrine systems.

A surprising chemical additive is citric acid. It sounds like a tangy substance squeezed out of a fresh lemon. But the citric acid used by food manufacturers comes from a very different source. Nearly all is made in factories using a fermentation process with the mold *Aspergillus niger*, mainly in China. The problem is that traces of the mold can end up in the finished product, potentially triggering allergic or autoimmune reactions.[29] In a recent research study, citric acid reliably triggered migraines in one of our participants, and others have blamed it for a wide range of symptoms. Just how risky it may be is mostly a matter of speculation. Researchers and product manufacturers have assumed it is safe without much in the way of testing. If you check food and beverage labels, you will be surprised at how frequently it crops up.

You can be exposed to chemicals without eating or drinking *anything*. Triclosan, for example, is an antibacterial that until recently was used in many hand soaps, detergents, and other products. Because some evidence suggests health risks, including disrupted thyroid function, most commercial products no longer include it. Check their labels to be sure. Colgate Total toothpaste *does* include it and boasts about triclosan's ability to prevent gingivitis. It may indeed work, but it's an open question as to whether good hygiene depends on added chemical exposures.

As we have seen, pesticides and industrial chemicals often end up in rivers and oceans, which are increasingly becoming a massive human sewer. Fish ingest these pollutants, which typically concentrate in their body fat. And because fish are carnivorous (unlike humans' terrestrial prey, who are mainly vegetarian), pollutants concentrate as smaller fish are eaten by bigger ones. Similarly, chemicals end up in cows' udders, concentrating in milk fat. In your body, these chemicals often end up in body fat, including breast tissue, which ends up as a chemical reservoir.

When you see white strips of fat in a steak at the meat counter or hear someone speak of "fatty fish," like salmon, think about the chemicals that may have accumulated in that stored fat. The same is true of your own body fat, which can gradually dispense its accumulated toxins into your bloodstream. A study of 431 overweight women and men found that those with the highest levels of chemical pollutants in their bloodstreams were most likely to have elements of metabolic syndrome—the combination of problems with body weight, blood pressure, blood sugar, and blood lipids mentioned above.

Hormones in Meat and Dairy Production

One final source of unwanted chemical exposures comes from animal products. In the United States and some other countries, hormones are routinely used in meat and dairy production. For beef, six different hormones are used, including testosterone, estradiol, progesterone, and three synthetic hormones—zeranol, melengestrol acetate, and trenbolone acetate. They are implanted as small pellets on the back of the ear in hopes of getting more meat with less feed. A relative of zeranol, called zearalenone, is not part of an implant. Rather, it is produced by fungi and ends up in meat as a result of contaminated feed. It has also been found in contaminated popcorn.

While the meat industry holds that the hormone traces in meat are too small to do any harm, researchers at the Rutgers Cancer Institute of New Jersey found reason for concern. They took urine samples from 163 girls, aged nine to ten. Zeranol showed up in 20 percent of the girls, and zearalenone appeared in 55 percent. Girls with detectable amounts of these compounds were significantly shorter than other girls.[30]

Some dairy farmers use genetically engineered bovine growth hormone (also called *bovine somatotropin*) to boost milk production.

The injections increase the incidence of mastitis—inflammation of the udder, which is usually treated with antibiotics. The drug is banned in Canada, the European Union, Australia, New Zealand, Japan, and Israel, but used on some U.S. farms.

The way to avoid these exposures is to steer clear of animal products. While some consumers seek out meat and dairy products produced without these hormones, they are still subject to the hormone-altering effects of meat and dairy products themselves.

Switching to a cleaner diet—avoiding animal products and choosing chemical-free foods as much as possible—will help you reduce the chemical traces in your body. Researchers tested the breast milk of women who had adopted vegan diets, finding that environmental contaminant levels were far lower than those for meat-eaters.[31]

But it takes time. Over the short run, a diet change that causes you to melt off some body fat may temporarily *increase* the chemicals in your bloodstream.[32] But evidence suggests that, over time, your body will gradually rid itself of most common pollutants.

Jeanne

Jeanne is a chemistry teacher at a private school in New York. As a teacher, she was at the top of her game, but physically, she was not well at all. "I was chronically fatigued," she said. "And over time, I gained a great deal of weight. My blood pressure was high, and I was constantly getting sick with colds, flu, bronchitis, and urinary tract infections."

She had grown up as one of five children eating a fairly typical American diet: breakfasts of eggs, toast, and cereal, and bacon or sausage on weekends. Lunch was whatever was on the school menu, and dinner was usually meat and two vegetables. Outside the home, she discovered chips, candy, fast food, popcorn with plenty of butter and salt, cheesy pizza, and chocolate. At holidays, overeating was the family norm. Weight problems plagued her family members.

"I have been on all the diet trends my entire life," she said. "Weight Watchers, Jenny Craig, Scarsdale, low-calorie—you name it, I have tried it. My weight has been up...then down...then up...then down."

Her turning point was a trip to the emergency room. With no apparent cause, she had a 105-degree fever and blood pressure through the roof. Doctors kept her in the hospital for a week. "I have never been so sick in my entire life," she said. "The good news was that the doctor who treated me was not only a medical doctor but was also a nutritionist. Beyond trying to track down the cause of the fever, she also kept asking me what food I ate. And the lightbulb went off." It was time for a change. Her doctor recommended a plant-based diet as a way to lose weight and regain her health, and gave her some books to help her get started. "Since that day, I haven't looked back."

Meat was easy to leave behind. She wasn't a fan of it anyway. She did have some trouble breaking up with Greek yogurt, cheese, and processed foods, but slowly and steadily, her diet improved.

Not long thereafter, her doctor checked her thyroid. It turned out that it was not doing its job. Her TSH level was 4.1 mU/L (high TSH levels indicate a sluggish thyroid, and a level above 4 suggests hypothyroidism). As she lost weight, her TSH climbed to 4.7, and her doctor put her on thyroid medication.

Was it that chemical traces in body fat had been released as she lost weight? Maybe. Many environmental toxins are fat-soluble. Ditto for chemicals in personal and home cleaning products. They accumulate in body fat, and as fat begins to break down, these toxins are released into the bloodstream.

In any case, she lightened up her diet, focusing on simpler foods: soups, salads, beans, rice, and fruits. She found she could make a mean berry mixture, with a base of oats, topped with lemon juice and zest. And as unwanted weight continued to drop away, and her TSH came back down, her doctor stopped her medications.

"I have lost a hundred pounds to date. My energy levels are off the charts. I never take naps anymore. They used to be a staple of my life. I rarely get sick; if I do, it is usually very short." For the past several years, her TSH has been firmly in the normal range— around 2—with no thyroid medication. "I was told that I would be on thyroid medication for the rest of my life. That turned out to be entirely wrong."

She noticed one other benefit. Her son was a competitive high school wrestler, and as the family menu shifted, his performance improved noticeably. While the other wrestlers fatigued early, his endurance stayed strong, especially during weekend-long tournaments. During these marathon events, his competitors ate pepperoni pizza, cheese pasta, chips, candy, and bagels with cream cheese. Her son ate out of his family's cooler, avoiding these processed foods. She noticed that as the tournaments progressed and the bouts became more intense and competitive, the other wrestlers' energy and strength waned, while her son's endurance and strength stayed high.

Choose Organic

You can reduce your chemical exposures by avoiding animal products. Needless to say, animals on farms are exposed to herbicides and other chemicals, which are concentrated in their flesh, as do fish.

When you buy produce, it helps to choose organic varieties. This matters more for some plants than others. Take spinach, for example. A spinach leaf is thin and fragile and is easy prey for insects passing by. So most spinach growers use pesticides routinely. But a sweet potato is different. It is underground. There are still some hungry little mouths in the garden, but it's nothing like it is for spinach. So pesticides are not used so much.

Also, let's say you were to eat a pineapple or an avocado. You would throw away the peel, along with whatever chemical residues may lie on the surface, unlike spinach, which has no peel, or apples or grapes, whose peel is commonly eaten.

With these considerations in mind, the Environmental Working Group publishes a handy list of fruits and vegetables where choosing organic makes a real difference, based on the most recent data from the U.S. Department of Agriculture.[33] These include:

Strawberries
Spinach
Kale
Nectarines
Apples
Grapes
Peaches
Cherries
Pears
Tomatoes
Celery
Potatoes
Hot peppers

Those for which chemical treatments are used less often and for which the organic-versus-conventional distinction is less important include:

Avocados
Sweet corn
Pineapples
Sweet peas

Onions
Papayas
Eggplant
Asparagus
Kiwi
Cabbage
Cauliflower
Cantaloupe
Broccoli
Mushrooms
Honeydew

These lists change from year to year, and I would encourage you to check EWG's website for the latest list.

A large study in France following 68,946 adults over five years found that those who chose organic foods most often were 25 percent less likely to develop cancer. Looking at specific cancer types, postmenopausal breast cancer risk was 34 percent lower among those eating more organic foods. Lymphoma risk was 76 percent lower, and non-Hodgkin's lymphoma risk was 86 percent lower.[34]

There is another advantage of supporting organic growers. It means fewer chemicals will end up in rivers and streams. Currently, six billion pounds of pesticides are used on food crops every year, worldwide. Favoring organic agriculture reduces that number and makes for a healthier environment all the way downstream.

By the way, organic produce is not necessarily more nutritious. That is, it does not have more vitamins or minerals. A comprehensive report from the Food Standards Agency in the UK found that nutritionally, organic and conventional produce were pretty much the same. Even so, it is much *cleaner*—more likely to be free of the chemicals you want to avoid.

One other advantage of organic produce: By law, products that are labeled "organic" cannot be genetically modified. Genetic modification means that food manufacturers have used viruses to smuggle genes into cells or have physically inserted genes into the nucleus in order to alter DNA—the fundamental blueprint that makes a plant or an animal what it is. They are trying to improve on the plants Mother Nature came up with. The FlavR SavR tomato, for example, was introduced in 1994 by Calgene (which later became a subsidiary of Monsanto), boasting that it stayed firm as it ripened, giving it a longer shelf life. You may have been perfectly happy with a normal tomato. But commerce thrives on products that can sit in warehouses and on store shelves for longer periods.

When I visit my childhood home in North Dakota, I am always impressed by the vast acreage of identical corn and soybean crops. These GMO (genetically modified organisms) plants extending as far as the eye can see are beautiful, but concerning. Of course, most are not grown for human consumption. They will fatten up cows, pigs, and chickens and are of particular concern to those who eat meat or dairy products, a habit I hope I am talking you out of as you read about their effects on your body.

How risky are genetically altered crops? It is really anyone's guess. The concerns are that their genes could somehow insinuate their way into your own DNA or into the bacteria that normally inhabit your digestive tract. Or perhaps a new genetically modified crop will trigger new allergies. In Brazil, soybeans were engineered to contain a gene taken from Brazil nuts. As a result, people with nut allergies became allergic to the new variety of soybeans.

Some believe (or hope) that genetically modified foods are safe.[35] On the other hand, their effects may be subtle—for example, readjusting hormone balance toward weight gain or causing cancer many years down the road—and may escape notice.[36] My own view is that genetically altered foods provide no health benefits, and while scientists argue about the risks, it makes sense to stay out of the experiment.

In the United States and Canada, there is no legal requirement that manufacturers disclose whether they have used genetic engineering. Nearly all animals raised for meat, dairy products, and eggs are fed genetically modified feed grains, and most cheeses are prepared using genetically engineered enzymes (rennet). At the moment, most of the corn, soybean, cotton, Hawaiian papaya, and canola crops in the United States are genetically modified, except those labeled organic, which are never GMO (e.g., organic soymilk or tofu is not GMO). Apples, oranges, bananas, broccoli, and most other fruits and vegetables are not GMO.

What about meat and dairy products that carry "organic" labels? Are they any better than "conventional" varieties? For products to be labeled "organic," the USDA requires that commonly used agricultural chemicals not be used or that farmers observe a washout period before meat or milk from treated animals can be sold. That is fair enough, but dairy products still contain the hormones *made by the cow*. And because these products have plenty of fat and no fiber at all, they cause hormonal disruptions of their own. Whether they're organic or not, you are better off without animal products. And, of course, "organic" does not mean that the animals were well treated.

Check the Label

The term *organic*, as used by government regulators, means that the use of pesticides and other chemicals is greatly restricted or avoided altogether. In the United States, the Department of Agriculture's National Organic Program sets the definitions. Produce that is grown in accordance with government standards can carry a "USDA Organic" label. For products with multiple ingredients (e.g., soups or baked goods), there are three different labels:

- "100% organic" indicates that the product contains only organic ingredients.

- "Organic" indicates that 95% of the product consists of organic ingredients.
- "Made with organic ingredients" indicates that at least 70% of the ingredients are organic.

If the produce you buy does not have a label, check the little sticker with the four-digit price look-up (PLU) code—the one that tells cashiers how to ring up fruits and vegetables. If the code is preceded by a "9," it is organic. By law, organic foods cannot be genetically modified. So even though the soybeans fed to chickens, cows, and pigs are typically GMO, the soybeans used to make your organic soymilk will not be.

Protect Yourself

While scientists and industry arm-wrestle over the risk of pesticides and other chemical exposures, my advice is to avoid them to the extent you can. After all, pesticides' whole *raison d'être* is to snuff out the lives of insects or other critters. It is a lot to ask that they be entirely safe for you and your little ones. Similarly, other commercial chemicals were designed for reasons other than your health, and many have unintended effects.

Let me provide some practical tips for protecting yourself and your family, beginning with two quick caveats:

First, we cannot eliminate all chemical exposures. Their use has been so widespread in agriculture, industry, and commercial manufacturing that traces will show up in food, water, and commercial products despite our best efforts.

Second, we cannot be entirely sure about the safety of chemicals. Those asserting the safety of chemicals have often relied on tests on rats and mice. Those indicting chemicals have called on animal tests as well. It has become clear that these tests are often crude and do not apply to humans. Moreover, their results can be manipulated depending on which animals are chosen, how the chemicals are administered, and how the results are interpreted. More helpful are studies investigating actual human exposures, as well as the new generation of human-cell-based test methods. Even so, these methods have their weaknesses, too; it is not always easy to gauge the safety of new chemicals.

With these cautions in mind, here are some simple steps that go a long way toward protecting us from chemical exposures:

1. Avoid animal products. Many environmental chemicals bioaccumulate in animal tissues.
2. Favor organic produce, especially for fruits and vegetables that are frequently pesticide-treated.
3. Favor fresh or frozen foods over canned. Some manufacturers (e.g., Eden Foods, Muir Glen) make a point of using BPA-free cans, as you will see on their labels.
4. Prepare beans from scratch. It is easy, amazingly inexpensive, and chemical-free.
5. In the microwave, use glass containers, not plastic.
6. If you buy plastic products, look for BPA-free labels or check the recycling number on the bottom, printed inside a triangle made of three arrows. A 3 or 7 indicates that the product may contain BPA, unless it is designated as BPA-free.
7. Drink clean water. Spring water beats tap water. If your spring water comes in plastic bottles, keep them away from heat. If you use tap water, filters can remove many pollutants. Check the owner's guide to ensure that your filter has been certified by NSF International (an independent rating firm) to meet American National Standards Institute's Standard 53 for reduction of volatile organic compounds. That will cover atrazine and many other pollutants.

8. Skip thermal receipts if you don't really need them, and wash your hands after touching them. If you handle receipts in your daily work, wear gloves.

9. Check labels on personal care products. The fewer additives, the better.

One final comment: Avoiding chemical exposures is a good idea. But this step should be taken *in addition to a healthful diet, not instead of it.*

Let us take a lesson from tobacco. When tobacco was shown to cause cancer, heart disease, and many other problems, the American Spirit company launched additive-free tobacco products, starting in 1982. The myth that cleaner tobacco is somehow less deadly than tobacco with additives was born. Surprising as it may sound, an enormous number of people believe it. While tobacco sales have plummeted overall, American Spirit is making a killing. But just as an additive-free bullet does not modify the outcome of Russian roulette, tobacco is tobacco: It causes cancer and other deadly problems regardless of what is or is not added to it.

The point is this: Unhealthful diets can harm you, regardless of whether they have been produced with extra chemicals. Partly, that is because potentially harmful chemicals are intrinsic to certain products. Cheese and other dairy products, for example, commonly contain estrogen traces that came from the cow. Also, because animal products contain unhealthful fats and have no fiber, they impair your natural hormone balance, as we have seen in detail.

So, it pays to start with a healthful plant-based diet, and then add chemical-reduction strategies to it, rather than focusing on chemicals alone.

Menus

Here are some simple menus using the recipes in this book. Of course, the meals and days can be interchanged as you like. We suggest that you begin on Saturday, with the option of preparing some foods in advance on the weekend.

WEEK 1

Day 1

Breakfast: Benedict Crostini with seasonal fresh fruit
Lunch: Cauliflower Buffalo Chowder with Green Bean & Potato Salad
Dinner: Butternut Pasta with Shiitake Miso Soup

Day 2

Breakfast: Butternut Breakfast Tacos
Lunch: Mediterranean Croquettes with Tzatziki Sauce and Bayou Quinoa
Dinner: Moroccan Pizzas with garden salad and Peach Vinaigrette

Day 3

Breakfast: Purple Power Smoothie with 1 ounce raw almonds
Lunch: Thai Crunch Salad with cubed extra-firm tofu
Dinner: Carrot & Pea Curry with Perfect Brown Rice

Day 4

Breakfast: Breakfast Pilaf with a banana or orange
Lunch: Kung Pao Lettuce Wraps with Brazilian Rice
Dinner: Middle Eastern Lentil Soup (Shorbat Adas) with Caesar Salad

Day 5

Breakfast: Italiano Scramble with whole-wheat toast
Lunch: Double Up Bella Burgers with Ruby Root Salad
Dinner: Shanghai Noodles with cubed extra-firm tofu and Arame
Salad

Day 6

Breakfast: Perfect Pancakes
Lunch: Mediterranean Croquettes with Green Bean & Potato Salad or
Middle Eastern Lentil Soup (Shorbat Adas)
Dinner: Southwest Lentil Mac with garden salad and Ranch Dressing

Day 7

Breakfast: Kale & Whatever Frittata with fresh berries
Lunch: Autumn Chowder or Chipotle Chili
Dinner: East Meets West Tacos with Curried Rice

WEEK 2

Day 8

Breakfast: Green Smoothie Muffins with 1 ounce almonds
Lunch: Butternut Pasta with garden salad and Strawberry Vinaigrette
Dinner: Double Up Bella Burgers with Chipotle Slaw and Pizza Pasta
Salad

Day 9

Breakfast: Baked sweet potato with fresh seasonal fruit
Lunch: Buffaloaded Pasta Salad with Chipotle Chili
Dinner: Roasted Quinoa Pie with Caesar Salad

Day 10

Breakfast: Breakfast Pilaf with 1 ounce raw almonds
Lunch: BBQ Bean Tortas with Citrus Sweet Potatoes
Dinner: Vegan Cauliflower Crust Pizza and Caesar Salad

Day 11

Breakfast: Granola with sliced banana and vegan yogurt
Lunch: Vegetable and avocado sushi with Grilled Corn Salad
Dinner: Mediterranean Croquettes with Tzatziki Sauce and Ruby Root
    Salad

Day 12

Breakfast: Green Smoothie Muffins with 1 ounce raw almonds
Lunch: Thai Crunch Salad with Thai Peanut Dressing
Dinner: Spaghetti with marinara sauce or Basic Polenta

Day 13

Breakfast: Strawberry Shortcake Polenta
Lunch: Mexican Quinoa with Chipotle Slaw and fat-free refried beans
Dinner: Coconut Cauliflower Curry with Perfect Brown Rice

Day 14

Breakfast: Oatmeal with fresh berries or sliced banana
Lunch: Simple Spring Rolls with Thai Peanut Dressing
Dinner: Mongolian Vegetable Stir-Fry with Perfect Brown Rice

Recipes

Breakfasts
Green Smoothie Muffins
Kale & Whatever Frittata
Purple Power Smoothie
Strawberry Shortcake Polenta
Perfect Pancakes
Benedict Crostini
Butternut Breakfast Tacos
Italiano Scramble
Breakfast Pilaf

Salads and Starters
Thai Crunch Salad
Buffaloaded Pasta Salad
Roasted Vegetable Salad
Arame Salad
Grilled Corn Salad
Thai Peanut Dressing
Avocado-Tahini Dressing
Basic Potato Salad

239

Coconut Dressing
Summer Panzanella Salad
Tabbouleh Spring Rolls
Caesar Salad
Pizza Pasta Salad
Green Bean & Potato Salad
Simple Spring Rolls
Ruby Root Salad

Soups

Savory Dill & Potato Soup
Cauliflower Buffalo Chowder
Middle Eastern Lentil Soup (Shorbat Adas)
Chipotle Chili
Coconut Cauliflower Curry
Autumn Chowder
Shiitake Miso Soup

Sandwiches and Wraps

Mediterranean Croquettes
Kung Pao Lettuce Wraps
BBQ Bean Tortas
Double Up Bella Burgers
Black-Eyed Pea Tacos
Rainbow Nori Rolls
Kale & Sweet Potato Sushi

Mains

Vegan Cauliflower Crust Pizza
Mongolian Vegetable Stir-Fry
Minestrone Polenta
Carrot & Pea Curry
East Meets West Tacos

Breakfasts

Green Smoothie Muffins
Makes 12 muffins

2 cups flour
½ cup brown or raw sugar
1 teaspoon baking soda
½ teaspoon ground cinnamon (optional)
Zest of 1 lemon
1½ cups (8 ounces) pineapple chunks (fresh or frozen)
2 cups fresh spinach
1–2 very ripe bananas
1 tablespoon vanilla extract
¼ cup plain soymilk or almond milk

Preheat oven to 375°F and line a muffin pan with paper liners or silicone cups, or use a nonstick pan.

In a large bowl, whisk together flour, sugar, baking soda, cinnamon (if using), and lemon zest. Set aside. In a blender, combine pineapple, spinach, bananas, vanilla, and milk. Blend until creamy. Pour wet mixture into dry ingredients and stir until just combined. If the batter is dry, add 2–4 tablespoons milk. Spoon batter evenly into muffin cups.

Bake 20–25 minutes, or until an inserted toothpick comes out clean. Let cool completely before serving.

Per muffin (¹⁄₁₂ of recipe): 124 calories, 3 g protein, 27 g carbohydrate, 10 g sugar, 0.5 g total fat, 2% calories from fat, 1 g fiber, 113 mg sodium

Kale & Whatever Frittata
Serves 4

2 cups any diced vegetables

1 cup chopped, stemmed kale

1 cup garbanzo bean flour

½ cup plain soymilk or almond milk

⅓ cup low-sodium vegetable broth (or water)

2–3 tablespoons chopped fresh herbs, such as basil, chives, or parsley

2 tablespoons nutritional yeast

1 tablespoon Dijon mustard

½ teaspoon spice, such as Italian seasoning, garlic powder, or onion powder

¼ teaspoon black salt (or plain salt)

Pinch paprika or cayenne pepper

Preheat oven to 450°F and place an oven-safe skillet in the oven while it heats. Pour ¼ cup water (or broth) into a separate skillet on medium-high heat. Add diced vegetables and cook until desired tenderness. Add kale and stir until bright green and excess liquid cooks off. Season with salt and pepper to taste.

Use an oven mitt to remove hot skillet from oven. Spray pan with nonstick cooking spray and use a clean paper towel to rub it around the surface. Carefully place vegetables in a single layer in bottom of skillet. In a bowl, mix remaining ingredients. Pour over vegetables, shaking the pan a few times so it settles evenly. Return to oven and bake 15–20 minutes. Broil for a few minutes for a crusty top. Remove from oven (it will look undercooked) and allow to cool slightly. Flip out of pan or cut into wedges.

Suggested vegetables: onions, peppers, tomatoes, zucchini, broccoli, mushrooms

Note: You can make "mini quiches" by mixing the kale and vegetables (or beans) into the batter and pouring batter into a muffin pan. Bake 10 minutes at 475°F, then reduce oven temperature to 450°F and bake 3–7 minutes more, or until tops are golden and firm to the touch.

Per serving (¼ of recipe): 140 calories, 9 g protein, 21 g carbohydrate, 5 g sugar, 3 g total fat, 16% calories from fat, 5 g fiber, 433 mg sodium

Purple Power Smoothie

Serves 1

1 cup plain soymilk or almond milk
2 dates (or 2 tablespoons raisins)
1 frozen ripe banana
2 cups frozen mixed berries (or other berries)
1–2 cups fresh spinach (optional)
Fresh mint or basil (optional)
1 tablespoon lemon juice (optional)

Combine soymilk or almond milk and dates in a blender and blend until well combined. Add remaining ingredients and blend, adding more soymilk or almond milk as desired, for consistency and thickness. Serve immediately.

Per serving: 460 calories, 11 g protein, 100 g carbohydrate, 61 g sugar, 6 g total fat, 11% calories from fat, 22 g fiber, 121 mg sodium

Strawberry Shortcake Polenta

Serves 3

1 (18-ounce) tube polenta, cut into ½-inch rounds
2 cups (10 ounces) frozen strawberries
5–6 ounces vegan yogurt (plain or vanilla)
1–2 tablespoons maple syrup or balsamic vinegar (optional)
Pinch ground cinnamon

To bake polenta (optional, since tube polenta is ready to eat and softer than baked polenta): Preheat oven to 350°F and line a baking sheet with parchment paper. Place polenta rounds on pan and bake 10–12 minutes, until golden in color and crisp (no need to flip).

Heat frozen strawberries in the microwave or on the stove until just warm. Slice, if desired. Top each polenta round with strawberries and a dollop of cold yogurt. Drizzle with maple syrup or balsamic vinegar (if using). Sprinkle with cinnamon.

Note: You can blend 10 ounces silken tofu with 1 ripe banana, 2 tablespoons fresh lemon juice, and 1–3 tablespoons soymilk (as needed) for a homemade yogurt. Add 1–2 tablespoons maple syrup or agave nectar for a sweetened yogurt.

Per serving (⅓ of recipe): 301 calories, 4 g protein, 71 g carbohydrate, 44 g sugar, 2 g total fat, 5% calories from fat, 5 g fiber, 235 mg sodium

Perfect Pancakes
Serves 4

1 cup flour
1 tablespoon baking powder
½ teaspoon ground cinnamon
1 cup plain soymilk or almond milk
2 tablespoons pure maple syrup
1 banana, sliced (or blueberries)
Pure maple syrup, for serving (optional)

Whisk flour, baking powder, and cinnamon in medium bowl until well combined. Stir in milk and maple syrup (or substitute applesauce or pumpkin puree). Let batter rest 10 minutes.

Meanwhile, heat a nonstick skillet or place parchment paper into a regular skillet (cut to size). Check batter. If it's very thick and heavy, add more milk as needed to thin. Pour ¼ cup of batter for each pancake. When bubbles appear, slide a spatula underneath and gently flip. Add banana slices (or blueberries) on top. Cook another 2–3 minutes. Repeat with remaining batter. Serve pancakes with additional maple syrup, if desired.

Per serving (¼ of recipe): 194 calories, 5 g protein, 41 g carbohydrate, 12 g sugar, 1 g total fat, 6% calories from fat, 2 g fiber, 396 mg sodium

Benedict Crostini

Serves 2

Mushroom Caps:

½ cup low-sodium vegetable broth
¼ cup balsamic vinegar
3 tablespoons minced onion (or shallot)
1 minced garlic clove
2 sliced portabella mushroom caps, stems removed

Hollandaise Sauce:

½ cup plain soymilk or almond milk
1½ teaspoons cornstarch
1 tablespoon nutritional yeast
Pinch ground turmeric
Pinch cayenne pepper
1½ teaspoons fresh lemon juice
1½ teaspoons vegan mayo (or plain vegan yogurt)

1 bagel, halved
1 sliced tomato
2 cups fresh spinach

For the Mushroom Caps: Combine broth, balsamic vinegar, onion or shallot, garlic, and salt and pepper to taste in large plastic bag or resealable plastic container. Add mushroom slices and marinate for about 30 minutes.

Meanwhile, for the Hollandaise Sauce: Heat milk in a saucepan over medium heat. Whisk cornstarch with 1 tablespoon water. Add to milk and bring to an almost boil. Cover, reduce heat as necessary to prevent boiling, and cook 2 minutes. Remove from heat and whisk in nutritional yeast, turmeric, cayenne, lemon juice, and mayo. Season with salt and pepper to taste.

Cook mushrooms in a nonstick skillet over medium-high heat for 3–4 minutes per side. Toast bagel halves and place on plates. Layer mushrooms, tomato slices, and spinach on each half. Spoon warm Hollandaise Sauce to top.

Note: For homemade vegan mayo recipe, see Tofu Mayo, page 292.

Per serving (½ of recipe): 273 calories, 14 g protein, 49 g carbohydrate, 16 g sugar, 3 g total fat, 11% calories from fat, 5 g fiber, 650 mg sodium

Butternut Breakfast Tacos
Serves 4

3 cups (½-inch) cubed peeled butternut squash (or sweet potatoes or regular potatoes)
1 tablespoon taco seasoning
8 (6-inch) corn tortillas
½ cup fresh salsa
1 cup guacamole
½ cup sliced green onions
Chopped fresh cilantro (optional)
Hot sauce (optional)

Preheat oven to 375°F and line a baking sheet with parchment paper. Rinse butternut squash cubes (or cubed potato) under cold water and shake off excess. Toss with taco seasoning until coated. Arrange in a single layer on pan and roast 30–40 minutes, or until fork-tender and browning, flipping halfway.

Spoon cooked squash (or potatoes) into corn tortillas. Top with salsa, guacamole, onions, and cilantro (if using). Drizzle with hot sauce if desired and serve.

Note: To cut up a whole butternut squash, first slice off both ends. Peel the squash. Cut in half crosswise, separating the neck from the body (where the seeds are). Cut the neck into ½-inch slices, then dice the slices. Cut the body in half lengthwise, then use a spoon to scrape out the seeds. Cut each half into slices, then dice.

Note: For heartier tacos, you can add refried beans or black beans. Try a different salsa such as salsa verde or a tropical salsa for a variation. Use Old Bay seasoning instead of taco seasoning (omit salsa) and hummus instead of guacamole for another savory option.

Easy Homemade Guacamole: Mash a ripe avocado. Stir in lime juice and ground cumin to taste. Add chopped onion and/or cilantro, if desired.

Per serving (¼ of recipe): 210 calories, 5 g protein, 39 g carbohydrate, 5 g sugar, 6 g total fat, 23% calories from fat, 9 g fiber, 518 mg sodium

Italiano Scramble

Serves 4

1–4 minced garlic cloves
1 cup diced tomato
1 (15-ounce) package firm tofu, drained and cubed
3 tablespoons nutritional yeast
1 tablespoon Italian seasoning
1 tablespoon Dijon mustard
1 teaspoon garlic powder
1 teaspoon onion powder
¼ teaspoon ground turmeric
¼ teaspoon black salt (optional)
Pinch cayenne pepper (optional)
Fresh basil and balsamic vinegar, to serve (optional)

Pour ¼ cup water (or broth) into a large skillet over medium-high heat. Add garlic and tomato, and cook until liquid evaporates. Add tofu and remaining ingredients, stirring to combine. Reduce heat to medium and cook and stir until heated through. (Add a splash of broth or soymilk if it starts to stick.) Season with salt and pepper to taste, plus more Italian seasoning, black salt, or cayenne pepper, if desired. Garnish with fresh basil and a drizzle of balsamic vinegar, if desired, and serve.

Variations: Add diced bell peppers and/or chopped spinach with the tomato. Substitute 2 cubed cooked potatoes, chickpeas, or navy beans for the tofu for a soy-free option.

Note: Black salt gives the dish an "eggy" taste.

Per serving (¼ of recipe): 109 calories, 12 g protein, 8 g carbohydrate, 2 g sugar, 5 g total fat, 36% calories from fat, 3 g fiber, 253 mg sodium

Breakfast Pilaf

Serves 2

¼ cup uncooked quinoa
¼ cup raisins (or dried cranberries)
1 sweet potato
2 cups fresh spinach
2 tablespoons plain vegan yogurt

Bring quinoa, ¾ cup water (or vegetable broth), and raisins to a boil in small saucepan. Reduce heat to low, cover, and simmer until quinoa is cooked and soft and raisins are plump, about 15 minutes.

Meanwhile, microwave sweet potato until tender, then dice. Turn heat off quinoa and stir in spinach until just bright green. Stir in diced sweet potato. Season with salt and pepper (or other seasonings) to taste. Serve with a dollop of vegan yogurt.

Note: You can substitute ½–¾ cup leftover cooked rice or another grain such as couscous.

Variations: Substitute Italian seasoning and sundried tomatoes for a variation. Add seasonings like Chinese five-spice or garam masala for a more savory breakfast.

Per serving (½ of recipe): 201 calories, 6 g protein, 42 g carbohydrate, 17 g sugar, 2 g total fat, 8% calories from fat, 5 g fiber, 119 mg sodium

Salads and Starters

Thai Crunch Salad
Serves 6

4–5 cups shredded kale and cabbage mix
½ cup thinly sliced red bell pepper
½ cup shredded carrots
½ cup sliced green onions
¼ cup fresh cilantro (optional)
½ cup shelled edamame (or chopped peanuts)
½ cup Thai Peanut Dressing (see page 254) or Coconut Dressing
(see page 256)

In a large bowl, toss together all ingredients except dressing. Add dressing and toss again.

Note: Add fresh mango and cucumber (diced or thinly sliced) for a variation. You can also substitute fresh mint (5–10 leaves, thinly sliced) for the cilantro. Use shelled edamame for a nut-free option.

Per serving (⅙ of recipe): 78 calories, 4 g protein, 9 g carbohydrate, 5 g sugar, 4 g total fat, 39% calories from fat, 2 g fiber, 152 mg sodium

Buffaloaded Pasta Salad
Serves 4

Ranch Dressing:

¼ cup plain soymilk or almond milk
1½ teaspoons fresh lemon juice
3 tablespoons plain hummus
¾ teaspoon Dijon mustard
½ teaspoon dried dill
¼ teaspoon garlic powder
⅛ teaspoon onion powder
1½–2 tablespoons Frank's Hot Sauce

2 cups uncooked pasta, any small shape
1 cup diced celery
1 cup chopped cauliflower florets
1 cup shredded carrots

For the Ranch Dressing: In a small bowl, combine milk and lemon juice. In another bowl, whisk together hummus, Dijon, and spices. Slowly whisk in milk mixture, adding only enough to make a creamy dressing. Add more dill or garlic to taste. Stir in Frank's Hot Sauce to taste. Chill until serving.

Cook pasta as directed on package. Drain, rinse under cool water, and return to pot.

Meanwhile, pour ¼ cup water (or broth) into a large skillet over medium-high heat. Add celery and cauliflower and cook 5 minutes. Reduce heat to medium, cover, and cook until vegetables are fork-tender. Add to cooked pasta with carrots. Add Ranch Dressing and toss well.

Variation: Substitute 2–3 cups cooked rice or quinoa (or other grain), and add chickpeas.

Per serving (¼ of recipe): 332 calories, 13 g protein, 63 g carbohydrate, 5 g sugar, 3 g total fat, 8% calories from fat, 6 g fiber, 269 mg sodium

Roasted Vegetable Salad

Serves 4–6

2 diced large white or yellow potatoes
8 ounces cherry tomatoes
¼ cup diced red onion (optional)
1 (15-ounce) can low-sodium chickpeas, drained and rinsed
1–2 teaspoons lemon zest
2 tablespoons capers or olives (any kind) (optional)
¼ cup balsamic vinegar
4–8 cups arugula (or other lettuce)

Preheat oven to 400°F and line three baking sheets with parchment paper. Place potatoes, cherry tomatoes, and onion (if using) in a single layer on two of the pans (make sure the potatoes don't touch). Sprinkle potatoes with salt and pepper. Toss chickpeas with lemon zest plus salt and pepper. Arrange in a single layer on third baking sheet. Place all the pans in the oven and roast 25–30 minutes, turning potatoes once and chickpeas twice. Tomatoes should be shriveled, chickpeas crisp (but not hard), and potatoes tender and golden. Five minutes before chickpeas are done, stir in capers, if using.

Toss roasted vegetables together in large bowl. Drizzle with balsamic vinegar and toss again. Add additional capers or lemon zest (or lemon juice) if desired. Serve over arugula.

Note: Add 1–2 pears or apples, cored and diced, for a variation. You can also substitute cubed tofu for the chickpeas, baking on parchment paper until golden, turning a few times, or use 1½ cups cooked lentils (do not bake the lentils).

Per serving (¼ of recipe): 262 calories, 10 g protein, 52 g carbohydrate, 8 g sugar, 2 g total fat, 7% calories from fat, 8 g fiber, 328 mg sodium

Arame Salad

Serves 4

1 cup arame (seaweed)
1 cup water
1 cucumber, peeled
½ teaspoon salt
2 tablespoons lemon juice
1 tablespoon rice vinegar
1 teaspoon low-sodium soy sauce
2 tablespoons water
12 leaves butter lettuce

Soak the arame in 1 cup of water for 15 minutes, until it is soft. Meanwhile, cut the cucumber lengthwise, scoop out the seeds, and slice into thin crescents. Spread them on a plate, sprinkle with salt, and put them in a bowl and let them stand for 15 minutes. Drain thoroughly.

Drain the arame and add to the bowl of cucumber slices and stir in the lemon juice, vinegar, soy sauce, and water. Place 3 lettuce leaves on each plate, and top with the arame-cucumber mixture.

Per serving (¼ of recipe): 17 calories, 2 g protein, 4 g carbohydrate, 1 g sugar, 0.2 g total fat, 10% calories from fat, 2 g fiber, 376 mg sodium

(Recipe by Neal Barnard)

Grilled Corn Salad

Serves 4

2–3 ears corn
2 cups diced, seeded tomatoes
1 (15-ounce) can low-sodium black, pinto, or kidney beans, drained and rinsed
2 tablespoons minced green chiles
¼ cup diced red onion
1–2 limes, zested and juiced
Chili powder or cayenne pepper (optional)
Hot sauce (optional)
½ avocado, diced

(continued)

To grill corn, remove husks and silk strands from corn. Place directly over heat on preheated grill. Rotate when dark spots appear and kernels deepen in color, 2–3 minutes. Continue grilling and rotating until ears are uniformly cooked, 10–12 minutes. Set aside to cool. Once cool, use a knife to cut kernels off the cobs.

In a large bowl, toss together grilled corn kernels, tomatoes, beans, chiles, and red onion. Add juice from 1 lime and toss again. Add more lime juice to taste. (Add zest for a stronger lime flavor.) If desired, sprinkle with chili powder or cayenne pepper for a kick, or drizzle with hot sauce. Stir in avocado before serving.

Note: For a spicy salad, add minced jalapeño. You can also add chopped cilantro or cooked grains, such as rice or quinoa. For a shortcut option, use frozen roasted corn (cook as directed on the bag).

Per serving (¼ of recipe): 203 calories, 9 g protein, 37 g carbohydrate, 6 g sugar, 4 g total fat, 17% calories from fat, 11 g fiber, 24 mg sodium

Thai Peanut Dressing
Makes 1¼ cups

¼ cup peanut butter
⅓ cup *hot* water
¼ cup lite coconut milk (or soymilk)
¼ cup sweet red chili sauce
2–3 tablespoons low-sodium soy sauce
1 tablespoon plus 1 teaspoon rice vinegar
¼ teaspoon garlic powder
¼ teaspoon ground ginger
Sriracha to taste

In a small bowl, whisk together peanut butter and hot water. Stir in remaining ingredients. Add more garlic powder, ground ginger, or sriracha to taste.

Per tablespoon: 29 calories, 1 g protein, 3 g carbohydrate, 2 g sugar, 2 g total fat, 53% calories from fat, 0 g fiber, 102 mg sodium

Avocado-Tahini Dressing

Makes 1 cup

1 small avocado
3 tablespoons tahini
3 tablespoons fresh lemon juice
1 teaspoon garlic powder
Ground cumin or cayenne pepper (optional)
Agave nectar, maple syrup, or date syrup (optional)

Combine all ingredients in a blender with 1–2 tablespoons water (or broth). Blend, adding more lemon juice as needed to achieve desired consistency. Taste, adding cumin or cayenne, if desired. Sweeten with agave nectar, maple syrup, or date syrup, if using.

Per tablespoon: 29 calories, 1 g protein, 1 g carbohydrate, 0 g sugar, 3 g total fat, 73% calories from fat, 1 g fiber, 4 mg sodium

Basic Potato Salad

Serves 2

1 pound red potatoes, cut into 1-inch pieces
¼ cup vegan mayo (or plain vegan yogurt)
1 tablespoon apple cider vinegar (or lemon juice)
1–2 teaspoons Dijon mustard
1 cup thinly sliced celery
1–2 tablespoons thinly sliced fresh chives (optional)
1 lemon

Boil potatoes in a pot of boiling water until just fork-tender, then let cool. In a small bowl, stir together mayo, vinegar, and mustard. In a large bowl, toss potatoes and celery with dressing, stirring to coat. Taste, adding more vinegar, lemon juice (or zest for a strong lemon taste), or mustard, if desired, plus salt and black pepper. Garnish with fresh chives, if desired.

Note: For homemade vegan mayo recipe, see Tofu Mayo, page 292.

Per serving (½ of recipe): 259 calories, 6 g protein, 44 g carbohydrate, 4 g sugar, 7 g total fat, 25% calories from fat, 7 g fiber, 491 mg sodium

Coconut Dressing
Makes ½ cup

⅓ cup lite coconut milk
2 teaspoons red curry paste
2–3 teaspoons low-sodium soy sauce
1–2 teaspoons agave nectar (or sugar)
½–1 teaspoon sriracha
Soymilk or lite coconut milk

In a small bowl, whisk together all ingredients. Thin with soymilk, additional coconut milk, or water, if desired.

Note: Substitute green curry paste for a variation.

Per tablespoon: 11 calories, 0 g protein, 1 g carbohydrate, 1 g sugar, 1 g total fat, 47% calories from fat, 0 g fiber, 104 mg sodium

Summer Panzanella Salad
Serves 6

1 cup cubed stale or toasted bread
1 (15-ounce) can low-sodium white beans, drained and rinsed
2 thinly sliced, pitted peaches
¼ cup thinly sliced red onion
¼ cup chopped fresh basil

Peach Vinaigrette:
2–3 tablespoons apple cider vinegar
1 tablespoon Dijon mustard
1 tablespoon peach preserves

Leave bread out to stale or toast, then cut into cubes. Toss with salad ingredients in large bowl.

For the Peach Vinaigrette: Whisk all ingredients in a small bowl. Taste, adding more vinegar, mustard, or preserves as desired. Add dressing to salad and toss again. Serve salad over mixed greens, if desired.

Note: Use balsamic vinegar and other preserves (or fresh fruit) for a variation. For example, for a Strawberry Vinaigrette, use strawberry jam and balsamic vinegar and substitute strawberries for the peaches in the salad.

Summer Wheat Berry Salad: Substitute 1 cup cooled cooked wheat berries for the bread cubes.

Per serving (⅙ of recipe): 117 calories, 6 g protein, 23 g carbohydrate, 7 g sugar, 1 g total fat, 5% calories from fat, 4 g fiber, 93 mg sodium

Tabbouleh Spring Rolls

Makes 6 rolls

Tahini Dipping Sauce:

¼ cup tahini

½ cup water (or broth)

2–3 tablespoons fresh lemon juice

1–2 teaspoons Dijon mustard

½ teaspoon garlic powder

¼ teaspoon onion powder

Tabbouleh Spring Rolls:

1 (16-ounce) bag broccoli slaw

¼ cup chopped fresh mint or parsley

1 cup diced tomato

1 minced garlic clove

Juice of 1 lemon

1 tablespoon nutritional yeast

½ teaspoon ground coriander (optional)

6 rice paper wrappers

For the Tahini Dipping Sauce: Whisk all ingredients in a bowl until combined. Add salt and pepper to taste. Thin with additional water (or broth) as needed. Refrigerate until ready to serve.

For the Tabbouleh Spring Rolls: Chop broccoli slaw until it resembles crumbles or pulse several times in a food processor. Transfer to a bowl and stir in remaining ingredients (except rice wrappers). Season with salt and pepper to taste.

Select a pan large enough for the rice paper to lie flat inside. Add 1 cup of very warm water to pan. Working with 1 sheet of rice paper at a time, soak for 20 seconds, or until it's pliable but not "gummy bear" soft.

(continued)

Place soaked wrapper on a cutting board and rub gently with wet hands to flatten. Place ¼ cup filling slightly below the center. Pick up edge closest to you and fold the rice paper up and over the filling mound. Fold right and left sides toward center. Pull the spring roll gently toward you as you roll it up burrito-style. Repeat with remaining wrappers and filling. Cover spring rolls with plastic wrap and refrigerate. (You can place parchment paper between them to prevent sticking.) Serve with Tahini Dipping Sauce.

Note: If your store does not sell broccoli slaw, you can chop raw broccoli florets and mix in shredded carrots.

Per spring roll (⅙ of recipe): 150 calories, 6 g protein, 21 g carbohydrate, 3 g sugar, 6 g total fat, 34% calories from fat, 4 g fiber, 400 mg sodium

Caesar Salad

Serves 1

Caesar Dressing:
2 tablespoons plain vegan yogurt (or vegan mayo)
½ teaspoon fresh lemon juice
½ teaspoon Dijon mustard
¼ teaspoon vegan Worcestershire sauce (or tamari)
Pinch garlic powder

3 cups chopped romaine lettuce
½ cup croutons
½ cup cherry tomatoes (optional)

For the Caesar Dressing: In a bowl, whisk together all ingredients until combined, adding garlic powder to taste. Chill until serving.

Toss lettuce, croutons, and tomatoes (if desired) in a large bowl. Add Caesar Dressing and toss again.

Note: To make croutons, toast 1 slice bread (or use stale bread) and toss in a bag with garlic powder and salt until lightly coated.

Note: For homemade vegan mayo recipe, see Tofu Mayo, page 292.

Per serving: 112 calories, 5 g protein, 20 g carbohydrate, 4 g sugar, 2 g total fat, 17% calories from fat, 4 g fiber, 194 mg sodium

Pizza Pasta Salad
Serves 2

2 cups uncooked pasta, any small shape (about 4 ounces)
1 cup diced tomato
1 cup diced bell pepper
¼ cup sliced black olives
¼ cup diced red onion (optional)
1 cup sliced mushrooms
½ teaspoon Italian seasoning (or dried oregano)

Italian Dressing:

3 tablespoons apple cider vinegar
¾ teaspoon Dijon mustard
¾ teaspoon Italian seasoning
¼ teaspoon onion powder
¼ teaspoon garlic powder
1 tablespoon nutritional yeast (or vegan parmesan)
1 tablespoon agave nectar

Cook pasta as directed on package. Drain, rinse under cool water, and place in a large bowl.

Meanwhile, chop vegetables while pasta cooks. Pour ¼ cup water (or broth) into a large skillet over medium-high heat. Add mushrooms and cook 3–4 minutes.

For the Italian Dressing: Whisk all ingredients in a small bowl.

In a large bowl, toss together pasta, dressing, vegetables, and Italian seasoning. Season with salt and pepper to taste.

Note: For homemade vegan parmesan recipe, see Vegan Parmesan, page 292.

Per serving (½ of recipe): 679 calories, 25 g protein, 130 g carbohydrate, 14 g sugar, 6 g total fat, 7% calories from fat, 11 g fiber, 183 mg sodium

Green Bean & Potato Salad
Serves 6

¼ cup chopped pecans
1–2 pounds red or russet potatoes
½–1 pound green beans, trimmed and cut in half
1 cup plain vegan yogurt (or vegan mayo)
⅓ cup red wine vinegar
Agave nectar (optional, to taste)
2 cups sliced cherry tomatoes
1–2 cups sliced grapes
½ cup diced red onion
Vegan bacon bits, for garnish (optional)

To toast pecans (optional): Preheat oven to 350°F. Spread pecans in a single layer on a baking sheet and bake 5–7 minutes or until lightly toasted. Meanwhile, boil potatoes in a pot of boiling water until just fork-tender, then cube. Let potatoes cool.

Blanch green beans in boiling water for 1–2 minutes. Drain and rinse under cold water.

In a small bowl, whisk together yogurt, vinegar, and a few drops of agave nectar, if desired.

In a large bowl, toss potatoes and green beans with yogurt dressing, stirring to coat. Fold in pecans, tomatoes, grapes, and red onion. Season with salt and pepper to taste.

Chill for at least 3 hours. Garnish with bacon bits, if desired.

Note: For homemade vegan mayo recipe, see Tofu Mayo, page 292.

Per serving (⅙ of recipe): 166 calories, 4 g protein, 29 g carbohydrate, 10 g sugar, 4 g total fat, 22% calories from fat, 4 g fiber, 162 mg sodium

Simple Spring Rolls

Makes 6 spring rolls

1 (14-ounce) bag coleslaw mix
1 cup thinly sliced red bell pepper
1¼ cups diced cucumber
6 rice paper wrappers
Sriracha (optional)
Thai Peanut Dressing, Coconut Dressing, or hoisin sauce (thinned
    with water), for dipping (optional)

In a large bowl, toss coleslaw mix, bell pepper, and cucumber
together.

Select a pan large enough for the rice paper to lie flat inside. Add
1 cup very warm water to pan. Working with 1 sheet of rice paper at
a time, soak for 20 seconds, or until it's pliable, but not "gummy bear"
soft. Place soaked wrapper on a cutting board and gently flatten. Place
vegetable mixture slightly below the center. Drizzle with sriracha, if
desired. Pick up edge closest to you and fold rice paper up and over
the mound of filling. Fold right and left sides toward center. Pull the
spring roll gently toward you as you roll it up like a burrito. Repeat with
remaining wrappers and filling.

Serve with Thai Peanut Dressing, Coconut Dressing, or hoisin sauce
(thinned with water), for dipping, if desired.

Note: Add fresh fruit like sliced peaches or avocado for a variation. For
more elaborate spring rolls, add shredded lettuce or carrots, thin strips
of tofu or mango, fresh basil, or soaked cellophane noodles to the filling.

Per spring roll (⅙ of recipe): 72 calories, 2 g protein, 16 g carbohydrate,
3 g sugar, 0.5 g total fat, 4% calories from fat, 2 g fiber, 57 mg sodium

Ruby Root Salad

Serves 4

3 beets

1 bunch chopped, stemmed kale (or other greens)

Juice of ½ lemon

¼ teaspoon salt

2 sliced pears

1½ cups cooked quinoa

1 (15-ounce) can low-sodium lentils, chickpeas, or navy beans, drained and rinsed

½ cup chopped red onion (optional)

¼ cup chopped walnuts

½ cup fat-free balsamic vinaigrette

To roast beets: Wrap beets in foil and bake in preheated oven at 375°F until fork-tender. Let cool, then rub skin off and dice.

Place kale in a bowl. Add lemon juice and salt and massage kale leaves. Top with individual rows or piles of the remaining ingredients. Drizzle with balsamic vinaigrette.

Note: Substitute grapefruit for the pears for a variation.

Per serving (¼ of recipe): 355 calories, 17 g protein, 60 g carbohydrate, 17 g sugar, 7 g total fat, 18% calories from fat, 13 g fiber, 522 mg sodium

Soups

Savory Dill & Potato Soup
Serves 2–3

1 quart low-sodium vegetable broth (4 cups)
½ cup diced onion
3 minced garlic cloves
1 cup sliced celery
1 cup cubed potatoes
2 cups shredded cabbage
1 (8-ounce) can low-sodium tomato sauce
2 tablespoons ketchup
2 tablespoons dry rolled oats
2 tablespoons fresh dill (or ¼ teaspoon dried dill)

Pour ¼ cup broth into a large pot over medium-high heat. Add onion and garlic, and cook until onion is translucent. Add celery and potatoes, and cook until almost tender, adding broth as needed. Add remaining broth, cabbage, tomato sauce, and ketchup, stirring to combine. Cover and bring to a boil. Reduce heat to low and simmer until vegetables are very tender. Add oats and dill, and cook 2–5 minutes more. Season to taste with salt and pepper, plus more dill, if desired.

Note: For a heartier soup, add 1 (15-ounce) can cannellini beans, drained and rinsed, with the tomato sauce. For a thick, creamy soup, blend part of the soup with an immersion blender or transfer 1–3 cups to a blender, blend, then stir back in.

Per serving (½ of recipe): 195 calories, 7 g protein, 43 g carbohydrate, 15 g sugar, 1 g total fat, 6% calories from fat, 7 g fiber, 627 mg sodium

Cauliflower Buffalo Chowder
Serves 3

2 cups low-sodium vegetable broth
½ cup chopped onion
1 minced garlic clove
2 cups diced carrots
1 cup sliced celery
2 cups cauliflower florets
1½ cups diced potatoes
1 (15-ounce) can low-sodium chickpeas, drained and rinsed
1 cup plain soymilk or almond milk
1–2 tablespoons Frank's Hot Sauce

Pour ¼ cup broth into a large pot over medium-high heat. Add onion and garlic, and cook until the onion is translucent. Add carrots and celery, and continue to cook until just fork-tender. Pour in remaining broth. Add cauliflower florets and potatoes. Cover and bring to a boil. Reduce heat to low and simmer until potatoes are very tender.

Transfer half of the soup to a blender and puree until creamy. Return to the pot (or use an immersion blender to partially blend the soup in the pot). Stir in chickpeas. Add soymilk or almond milk as needed to thin the chowder. Stir in Frank's Hot Sauce plus salt and pepper to taste. Drizzle with more hot sauce before serving, if desired.

Note: Use Frank's RedHot Original Cayenne Pepper Sauce, not Frank's wing sauce.

Per serving (⅓ of recipe): 286 calories, 13 g protein, 52 g carbohydrate, 14 g sugar, 4 g total fat, 13% calories from fat, 13 g fiber, 500 mg sodium

Middle Eastern Lentil Soup (Shorbat Adas)
Serves 2

1 quart low-sodium vegetable broth (4 cups)
½ cup finely diced onion
½ cup finely diced carrot
½ cup finely diced celery
2 minced garlic cloves
1 teaspoon ground cumin
½ teaspoon ground turmeric
¾ cup dry red lentils or yellow split peas
1–3 teaspoons fresh lemon juice
1–2 teaspoons fresh parsley

Pour ¼ cup broth into a large pot over medium-high heat. Add onion, carrot, celery, and garlic, and cook until vegetables are tender. Stir in cumin to coat, and cook 1–2 minutes. Add turmeric, stirring until it's yellow. Pour in remaining broth and lentils. Cover and bring to a boil. Reduce heat to low and simmer 15–20 minutes, or until lentils are cooked. Season with salt and pepper to taste. Add lemon juice just before serving. Garnish with parsley and serve with warm whole-wheat pita bread or vegan naan.

Per serving (½ of recipe): 294 calories, 20 g protein, 53 g carbohydrate, 7 g sugar, 2 g total fat, 4% calories from fat, 14 g fiber, 467 mg sodium

Chipotle Chili

Serves 3

½ cup diced onion

1 cup diced carrots

1 cup sliced celery

1 cup diced red or green bell pepper

3 minced garlic cloves

1 cup diced sweet potato

1–2 teaspoons chopped chipotle chiles in adobo sauce

2 teaspoons dried oregano

1 heaping tablespoon chili powder

1½ teaspoons ground cumin

1 (14-ounce) can fire-roasted diced tomatoes (undrained)

1 (15-ounce) can low-sodium black beans, kidney beans, or pinto
   beans (undrained)

1–2 tablespoons vegan Worcestershire sauce (optional)

3–4 sliced green onions (optional)

Chopped avocado (optional)

Plain vegan yogurt (optional)

Pour ¼ cup water (or broth) into a large pot over medium-high heat. Add onion, carrots, celery, peppers, garlic, and sweet potato. Cook until almost fork-tender. Stir in chipotle chiles (start with less if you have a milder palate), oregano, chili powder, and cumin. Cook for 1 minute. Add tomatoes (with their juices) and 2 cups water (or broth). Cover and bring to a boil. Reduce heat to low, uncover, and simmer for 45 minutes.

Stir in beans (drain first or add with their liquid for thicker chili) and cook over low heat until heated through. Thin chili with water (or broth) as needed. Stir in vegan Worcestershire sauce, if using, plus adobo sauce (from the can of chipotle chiles in adobo sauce) to taste, if desired. Garnish with green onions, avocado, and a dollop of tofu sour cream or plain vegan yogurt, if desired.

Note: Look for "chipotle chiles in adobo sauce," a canned item, in the international section of a grocery store.

Per serving (⅓ of recipe): 255 calories, 12 g protein, 52 g carbohydrate, 12 g sugar, 2 g total fat, 7% calories from fat, 19 g fiber, 388 mg sodium

Coconut Cauliflower Curry
Serves 6

¾ cup chopped onion

1 (14-ounce) can fire-roasted tomatoes (undrained)

2 cups diced sweet potatoes

1 tablespoon mild curry powder

4 cups cauliflower florets (about 16 ounces)

1 (15-ounce) can low-sodium kidney beans or chickpeas, drained
and rinsed

1 (14-ounce) can lite coconut milk

½ teaspoon salt

1 teaspoon ketchup or sugar (if needed)

2 cups cooked brown rice

Fresh cilantro (optional)

Pour ¼ cup water (or broth) into a large skillet over medium-high heat.
Add onion and cook until translucent. Stir in tomatoes (with juices from
the can). Use a spatula or fork to break up any large pieces. Add ½ cup
water (or broth) and sweet potatoes. Cover and bring to a boil. Reduce
heat to low and simmer until sweet potatoes are just fork-tender. Stir
in curry powder. Add cauliflower florets and simmer 15 minutes, until
cauliflower reaches desired tenderness and potatoes are very fork-
tender, adding ¼ cup water (or broth) as needed.

Stir in beans and coconut milk. Bring to an almost boil, then reduce
heat and simmer, uncovered, until heated through and liquid reduces
slightly. Season with salt. Taste. If it's acidic, add ketchup or sugar.
Add more curry powder to taste, if desired, and heat 5 minutes more
over low heat. Serve curry over cooked rice and garnish with cilantro, if
desired.

Note: You can substitute 1–2 cups peas for the beans in this recipe.

Per serving (⅙ of recipe): 254 calories, 9 g protein, 44 g carbohydrate,
9 g sugar, 6 g total fat, 20% calories from fat, 9 g fiber, 308 mg sodium

Autumn Chowder

Serves 3

¼ cup low-sodium vegetable broth
½ cup diced onion
8 ounces sliced white mushrooms
1–2 teaspoons rubbed sage (not powdered sage)
1–2 teaspoons dried thyme
2 cups plain soymilk or almond milk
2 tablespoons nutritional yeast
2 tablespoons low-sodium soy sauce
2 cups chopped, stemmed kale (or other greens)

Pour broth (or water) into a large pot over medium-high heat. Add onion and cook 5–7 minutes until translucent. Add mushrooms, sage, and thyme, and cook until mushrooms start to soften and release their juices. Add additional broth (or water) if needed to prevent sticking.

In a small bowl, whisk together milk, nutritional yeast, and soy sauce. Pour over mushrooms and stir well. Cook over low heat until heated through. Season to taste with pepper plus more sage or thyme, if desired. Add kale before serving and continue to cook over low heat, stirring occasionally, until kale is bright green.

Per serving (⅓ of recipe): 128 calories, 10 g protein, 17 g carbohydrate, 9 g sugar, 3 g total fat, 22% calories from fat, 4 g fiber, 476 mg sodium

Shiitake Miso Soup

Serves 4

4 ounces thin Asian noodles (or spaghetti)
1 quart low-sodium vegetable broth
8 ounces sliced shiitake mushrooms (or other mushrooms)
⅓ cup sliced green onions
2 minced garlic cloves
1 tablespoon minced ginger (about 1-inch root)
2 sliced carrots
2–3 tablespoons miso paste
1 tablespoon low-sodium soy sauce (optional)
2 cups spinach (or shredded cabbage)
Red pepper flakes or sriracha (optional)

Cook noodles as directed on package to al dente. Drain, rinse under cold water, and set aside.

Meanwhile, pour ¼ cup broth into a large pot. Add mushrooms and cook until just soft. Add green onions (save a little for garnish) with garlic and ginger and continue to cook until onions are translucent, adding a little water if needed. Stir in remaining broth and carrots. Cover, bring to a boil, then reduce heat and simmer until carrots are tender. Stir in cooked noodles and miso. Let stand for a few minutes, then taste, adding more miso (or soy sauce) as desired.

Add spinach (or cabbage) before serving, stirring until it softens. Add red pepper flakes or sriracha, if desired. Garnish with reserved green onions.

Note: To increase protein, add 1 cup cubed tofu. For a "fishy" flavored soup, add kelp or other seaweed a few minutes before serving.

Per serving (¼ of recipe): 171 calories, 7 g protein, 34 g carbohydrate, 4 g sugar, 1 g total fat, 6% calories from fat, 5 g fiber, 488 mg sodium

Sandwiches and Wraps

Mediterranean Croquettes
Serves 4

Tzatziki Sauce:

1½ cups diced cucumber
6 ounces plain vegan yogurt
1 minced garlic clove
2 teaspoons fresh lemon juice
¼ cup chopped fresh basil

Croquettes:

1½ cups cooked lentils
1½ cups cooked brown rice
¾ cup diced onion
2 minced garlic cloves
¼–½ cup minced fresh parsley
2 tablespoons nutritional yeast
1 tablespoon fresh lemon juice
1 teaspoon ground coriander
½ teaspoon baking powder
Cayenne pepper, to taste

For the Tzatziki Sauce: Mix all ingredients in a medium bowl. Refrigerate until ready to use.

For the Croquettes: Preheat oven to 400°F and line a baking sheet with parchment paper.

Place all ingredients in a food processor and process until well combined. Shape the mixture into about 12 small patties (croquettes) and place on the baking sheet.

Bake 7–10 minutes. Turn croquettes and bake 5–10 minutes more, or until crisp around the edges. For an appetizer, serve croquettes topped with a dollop of Tzatziki Sauce.

Note: Cook more lentils and brown rice than you need for this recipe so you can have extra to use in various recipes. For a delicious sandwich, stuff a pita or roll with croquettes, Tzatziki Sauce, lettuce, tomato, and cucumber. Leftover Tzatziki Sauce can be served with grilled or roasted vegetables.

Per serving (¼ of recipe): 251 calories, 13 g protein, 47 g carbohydrate, 5 g sugar, 2 g total fat, 8% calories from fat, 8 g fiber, 77 mg sodium

Kung Pao Lettuce Wraps
Serves 2

Kung Pao Sauce:
2 tablespoons low-sodium soy sauce
½–1 tablespoon agave nectar
½ tablespoon rice vinegar
¾ teaspoon cornstarch
¼–½ teaspoon garlic powder
Sriracha or cayenne pepper (optional)

Lettuce Wraps:
1 cup diced red bell pepper
⅔ cup sliced celery
1 (15-ounce) can low-sodium chickpeas, drained and rinsed
1 head lettuce, separated into lettuce cups
¼ cup crushed peanuts (optional)
Chopped fresh cilantro (optional)

For the Kung Pao Sauce: Whisk together all ingredients in a small bowl.

For the filling: Pour ¼ cup water (or broth) into a skillet over medium-high heat. Add bell pepper and celery and cook until tender. Add Kung Pao Sauce and chickpeas and mix well. Reduce heat to low and cook until sauce thickens to a light glaze.

Spoon filling into lettuce cups. Garnish with peanuts (if using) and cilantro (if using). Serve with additional sriracha, if desired.

Note: For a variation, add ¼ to ½ teaspoon each curry powder and ground cumin to the Kung Pao Sauce.

Per serving (½ of recipe): 279 calories, 14 g protein, 50 g carbohydrate, 18 g sugar, 4 g total fat, 13% calories from fat, 13 g fiber, 645 mg sodium

BBQ Bean Tortas

Serves 4

8 (6-inch) corn tortillas, or 4 hamburger buns, toasted
1 (15-ounce) can low-sodium black beans, drained and rinsed
½ cup barbecue sauce
½ cup chopped pineapple
1 cup fresh salsa (or diced tomato)
Guacamole (optional)

If using tortillas, preheat oven to 375°F. Bake corn tortillas for
5–10 minutes, or until crisp.

Warm beans with barbecue sauce. Spoon onto tortillas or buns and
top with pineapple, salsa, and guacamole, if desired.

Per serving (¼ of recipe): 293 calories, 9 g protein, 61 g carbohydrate,
18 g sugar, 2 g total fat, 6% calories from fat, 11 g fiber, 579 mg sodium

Double Up Bella Burgers

Serves 2

½ cup diced red onion
1 minced garlic clove
3 tablespoons balsamic vinegar
2 teaspoons Italian seasoning
2 large portabella mushroom caps, stems removed
2 burger buns (or 2 gluten-free tortillas)
1 cup spinach (or lettuce)
2–4 tomato slices

Secret Sauce:

2 tablespoons vegan mayo (or plain vegan yogurt)
½ teaspoon ketchup
½ teaspoon dill relish (or minced dill pickle)
¼ teaspoon white vinegar
½ teaspoon paprika
¼ teaspoon garlic powder
¼ teaspoon onion powder
Agave nectar (optional, to taste)

Pour ¼ cup water (or broth) into a skillet over medium-high heat. Add onion and garlic and cook for 2 minutes. Add balsamic vinegar, Italian seasoning, mushrooms, and ¼ cup water. Cover and bring to a boil. Reduce heat to low and simmer for 5 minutes. Turn mushrooms and cook 5 minutes more, or until mushrooms are tender, adding water (or broth) as needed to prevent sticking.

Meanwhile, prepare Secret Sauce: Combine all ingredients in a bowl and mix well. Add sweetener, if desired. Spread Secret Sauce on top half of buns. Place mushrooms, spinach, and tomato slices on bottom half on buns. Cover with top half of buns.

Note: For homemade vegan mayo recipe, see Tofu Mayo, page 292.

Per serving (½ of recipe): 227 calories, 9 g protein, 36 g carbohydrate, 11 g sugar, 6 g total fat, 22% calories from fat, 4 g fiber, 386 mg sodium

Black-Eyed Pea Tacos

Serves 2

2 cups diced sweet potatoes
½ teaspoon smoked paprika
½ cup diced red onion
1 cup diced green bell pepper
1 minced garlic clove
1 (15-ounce) can low-sodium black-eyed peas, drained and rinsed
½ teaspoon ground cumin
6 (6-inch) corn tortillas or taco shells
Avocado slices (optional)
Salsa (optional)
Chopped fresh cilantro (optional)
Lime juice (optional)
Hot sauce (optional)

Preheat oven to 400°F and line a baking sheet with parchment paper. Toss sweet potatoes with ½ teaspoon smoked paprika plus salt and pepper. Place in a single layer in the pan. Roast for 25 minutes, turning once, or until sweet potatoes are tender.

Pour ¼ cup water or vegetable broth into a skillet over medium-high heat. Add onion and bell pepper and cook until diced pepper is just tender. Add garlic and cook 1 minute more. Stir in black-eyed peas and cumin, plus more water or broth as needed. Cook over low heat, stirring, until heated through. Season with salt and pepper to taste.

Spoon bean filling and roasted sweet potatoes onto warmed tortillas or into taco shells. Top as desired with avocado, salsa, cilantro, lime juice, and hot sauce.

Note: For hard taco shells, drape each tortilla over two bars of your oven rack and bake at 375°F for 5–10 minutes, or until crisp.

Per serving (½ of recipe): 472 calories, 20 g protein, 95 g carbohydrate, 11 g sugar, 3 g total fat, 6% calories from fat, 16 g fiber, 368 mg sodium

Rainbow Nori Rolls
Makes 12 rolls

3 cups cooked rice (or quinoa)
2–4 tablespoons rice vinegar (optional)
Agave nectar (optional)
12 sheets nori seaweed
1 avocado, pitted, sliced, or mashed
1 cup diced red pepper
½ cup shredded beets
1 cup shredded cucumber (or zucchini)
¾ cup shredded carrots
Sesame seeds (optional)
Hoisin sauce (optional)

Season cooked rice (or quinoa) with the rice vinegar and a few drops of agave nectar if desired. Spread cooked rice (or quinoa) on top of each nori sheet and top with avocado and veggies, plus a sprinkling of sesame seeds if using. Roll tightly to close into a log shape. Repeat.

Optional: Cut nori rolls into smaller pieces. Dilute hoisin sauce with water for a dipping sauce, if desired.

Per roll (¹⁄₁₂ of recipe): 83 calories, 2 g protein, 15 g carbohydrate, 1 g sugar, 2 g total fat, 20% calories from fat, 3 g fiber, 27 mg sodium

Kale & Sweet Potato Sushi
Makes 12 rolls

2 cups cooked rice
2–4 tablespoons rice vinegar (optional)
2 sweet potatoes
4–6 garlic cloves, minced
1 bunch chopped, stemmed kale
1–2 tablespoons low-sodium soy sauce
12 sheets nori seaweed (or 4 tortillas)
4 green onions, sliced
1 avocado, pitted, sliced

Season cooked rice with 1–2 tablespoons rice vinegar if desired. Microwave sweet potatoes, then cube (or cube and roast until tender). Meanwhile, line a large skillet with ¼ cup water. Cook garlic for 2 minutes, then add kale, soy sauce, and 1–2 tablespoons rice vinegar if desired. Use tongs to stir and incorporate and mix, cooking until kale is bright green. (Set aside to cool slightly.) Lay nori sheets flat. Spread a thin, wide layer of rice across them and top with cooked sweet potato, kale, green onions, and avocado.

Gently roll with your hands. To seal the nori roll, dip your finger in water and moisten the end of the nori sheet, then press it against the roll to stick. Place roll, seam side down, onto a cutting board and cut into 4 pieces.

Per roll (¹/₁₂ of recipe): 80 calories, 2 g protein, 15 g carbohydrate, 2 g sugar, 2 g total fat, 20% calories from fat, 3 g fiber, 75 mg sodium

Mains

Vegan Cauliflower Crust Pizza
Serves 2–4

Crust:

2 tablespoons ground flaxseeds (or chia seeds)
1 (16-ounce) bag cauliflower rice
¼ cup flour (any)
2 tablespoons nutritional yeast
1 teaspoon garlic powder
¼ teaspoon onion powder (optional)
1½ teaspoons Italian seasoning (optional)
1–2 tablespoons tahini

Toppings:

½ cup tomato or marinara sauce
4–6 mushrooms, sliced
2 dry-pack sun-dried tomatoes, chopped
¼ cup sliced olives
½ cup pineapple, chopped
1 cup spinach
3–4 leaves fresh basil (optional)

For the Crust: In a small bowl, mix together the ground flax or chia with 3 tablespoons water. Refrigerate. Microwave cauliflower rice for 8 minutes or cook on the stove for 5 minutes. Once cool, place cooked cauliflower in a clean kitchen towel or cheesecloth and squeeze out all excess liquid. (Squeeze really hard! It needs to be super dry.) Transfer to a mixing bowl, add in refrigerated ground flax (or chia), and stir in remaining ingredients. Add 1–3 tablespoons water (as needed) to mix the dough (the drier the dough, the crispier the crust will be).

Preheat oven to 400°F. Turn a baking sheet facedown and line with parchment paper. Place dough ball in the center and cover with more parchment. Flatten and smooth into a 9-inch crust (about ¼ inch thick). Remove top piece of parchment (reserve) and bake 18–25 minutes, or until edges are slightly brown and crispy. Remove crust from oven and place parchment back on top. Put a baking sheet on top and turn over so the crust is on the new baking sheet. Remove top parchment (the old bottom piece) and discard.

For the Toppings: Spread a thin layer of tomato sauce on crust (you may not use the entire ½ cup). Add toppings except spinach and basil, if using, and bake 5–10 minutes more, or until toppings are warm. Add spinach and basil, if using, before serving (or bake and cook 1 minute more). Sprinkle pizza with additional nutritional yeast, if desired.

Per serving (½ of recipe): 313 calories, 15 g protein, 46 g carbohydrate, 17 g sugar, 11 g total fat, 29% calories from fat, 13 g fiber, 474 mg sodium

Mongolian Vegetable Stir-Fry
Serves 2

3 minced garlic cloves
1 tablespoon minced fresh ginger (about ½-inch piece)
Pinch red pepper flakes
3 green onions, sliced, divided into white and green parts
2 sliced portabella mushroom caps, stems removed
1 (14-ounce) bag frozen Asian stir-fry vegetables
1 tablespoon hoisin sauce
1–2 tablespoons low-sodium soy sauce
2 teaspoons cornstarch
2–3 cups cooked brown rice (or noodles)

Pour ¼ cup water (or broth) into a large skillet over medium-high heat. Add garlic, ginger, red pepper flakes, and white parts of green onions, and cook until onion is translucent. Add mushrooms and cook until tender. Add stir-fry vegetables and cook 1–2 minutes, or until heated through but still crisp.

In a small bowl, whisk ¼ cup cold water with hoisin, soy sauce, and cornstarch until smooth. Stir into vegetables. Reduce heat and allow to thicken. Serve over cooked rice or noodles. Garnish with reserved green onions.

Per serving (½ of recipe): 381 calories, 14 g protein, 78 g carbohydrate, 9 g sugar, 3 g total fat, 7% calories from fat, 10 g fiber, 450 mg sodium

Minestrone Polenta

Serves 2

¼ cup chopped onion
2–3 minced garlic cloves
1–2 teaspoons Italian seasoning
Pinch red pepper flakes
1 (14.5-ounce) can fire-roasted diced tomatoes (undrained)
¾ cup frozen mixed vegetables
1 teaspoon ketchup or sugar (optional)
1 (15-ounce) can low-sodium navy beans or chickpeas, drained and
    rinsed
1 cup polenta or grits
½ cup plain soymilk or almond milk

Pour ¼ cup water (or broth) into a medium saucepan over medium-high heat. Add onion and garlic, and cook until onion is translucent. Add Italian seasoning and a pinch of red pepper flakes, stirring to coat well. Add tomatoes (with juices), breaking up any large pieces with a spatula or fork. Cover and simmer 10 minutes, adding water (or broth) as needed to prevent burning or sticking. If desired, transfer half of the soup to a blender and puree. Return to the pan. Or, use an immersion blender to partially blend the soup in the pan.

In another saucepan, bring 2 cups water (or broth) to boil. Reduce heat to low and slowly whisk in polenta or grits. Whisk constantly to avoid clumps. Continue cooking and stirring until polenta is bubbly. Turn off heat and stir in soymilk or almond milk as needed for a creamy consistency. Season with salt and pepper to taste. Serve minestrone over polenta.

Per serving (½ of recipe): 611 calories, 23 g protein, 122 g carbohydrate, 12 g sugar, 4 g total fat, 5% calories from fat, 26 g fiber, 609 mg sodium

Carrot & Pea Curry

Serves 4

2–3 minced garlic cloves
2–3 tablespoons minced fresh ginger (about 1-inch piece)
1 tablespoon mild curry powder
1 teaspoon smoked paprika
1 (14-ounce) can lite coconut milk
2 tablespoons tomato paste
3 cups diced fresh carrots
2 cups frozen peas
¼ cup fresh cilantro, for garnish (optional)

Pour a thin layer of water (or broth) into a large pot over medium-high heat. Add garlic and ginger, and cook 1–2 minutes, or until just fragrant. Stir in curry powder and smoked paprika. Add coconut milk, tomato paste, and ½ cup water (or broth). Bring to a boil over medium heat while continuing to stir. Add carrots and stir. Cover and cook over low heat until fork-tender, about 15 minutes. Add peas and simmer, uncovered, for 5 minutes to allow the peas to cook and sauce to thicken. Season to taste with salt and garnish with cilantro, if using. Serve over cooked rice or with pita bread.

Note: For a more substantial curry, add cubed potatoes or cauliflower florets with the carrots. Or, add broccoli florets with the peas.

Per serving (¼ of recipe): 159 calories, 5 g protein, 22 g carbohydrate, 8 g sugar, 7 g total fat, 37% calories from fat, 7 g fiber, 254 mg sodium

East Meets West Tacos

Serves 4

1 tablespoon miso paste
1 tablespoon pure maple syrup
1 tablespoon rice vinegar
2 cups diced sweet potatoes
8 (6-inch) corn tortillas
½ cup guacamole
4 sliced green onions
Sriracha (optional)

Coconut Sauce:

¼ cup lite coconut milk
¼ cup fresh cilantro
¼ cup fresh basil
1 tablespoon fresh lime juice
⅛ teaspoon grated lime zest
⅛ teaspoon garlic powder

Preheat oven to 400°F and line a baking sheet with parchment paper. Whisk miso, maple syrup, and vinegar in large bowl. Add sweet potatoes and toss to coat. Spread in a single layer on pan. Bake 20–30 minutes, or until fork-tender.

Meanwhile, for the Coconut Sauce: Combine all ingredients in a blender and pulse until just combined.

Place sweet potatoes on tortillas. Top with guacamole, green onions, and sriracha, if desired. Drizzle with Coconut Sauce and serve.

Per serving (¼ of recipe): 216 calories, 5 g protein, 41 g carbohydrate, 8 g sugar, 5 g total fat, 18% calories from fat, 6 g fiber, 322 mg sodium

Moroccan Pizzas
Serves 4

Pizza Cheeze Sauce:

1 cup plain soymilk or almond milk
¼ cup nutritional yeast
¼ cup flour or cornstarch
1 tablespoon fresh lemon juice
1–2 teaspoons miso paste
1 teaspoon onion powder
1 teaspoon garlic powder
⅛ teaspoon ground mustard (optional)

½ teaspoon ground cumin
¼ teaspoon ground cinnamon
¼ teaspoon ras el hanout (Moroccan spice mix), plus more to taste
1 cup no-salt-added spaghetti sauce
4 pita breads
1 cup shredded carrots
1 cup sliced olives

For the Pizza Cheeze Sauce: Whisk all ingredients together in a saucepan. Bring just to a boil over medium heat. Reduce to low and cook, stirring frequently, until slightly thickened.

Preheat toaster oven or oven to 425°F. Stir cumin, cinnamon, and ras el hanout to taste (start with ¼ teaspoon) into marinara sauce. Spread spaghetti sauce and Pizza Cheeze Sauce on pitas. Top with carrots and olives. Bake 4–7 minutes, or until "cheeze" melts.

Per serving (¼ of recipe): 337 calories, 14 g protein, 57 g carbohydrate, 8 g sugar, 7 g total fat, 18% calories from fat, 7 g fiber, 678 mg sodium

Roasted Quinoa Pie

Serves 2

2 zucchini, sliced into half moons
½ cup diced red onion
1 cup diced red bell pepper
10 ounces cherry tomatoes, sliced
1½ cups cooked quinoa
1–2 tablespoons Italian seasoning
1½ cups no-salt-added spaghetti sauce

Preheat oven to 400°F and line a baking sheet with parchment paper. Place zucchini slices in a single layer on pan. Sprinkle onion and bell pepper over zucchini. Top with cherry tomatoes. Bake 25 minutes or until tomatoes are shriveled and vegetables are roasted and browned. Set aside.

Mix cooked quinoa with Italian seasoning (or add seasoning to cooking water when preparing quinoa). Stir in a spoonful or two of marinara sauce to coat. Place quinoa in the bottom of a small baking dish, creating a base layer. Spoon half the spaghetti sauce on top. Add roasted vegetables on top and season with salt and pepper. Bake 5–15 minutes, or until heated through and quinoa is toasted around the edges. Warm remaining spaghetti sauce and spoon over the top.

Per serving (½ of recipe): 347 calories, 13 g protein, 62 g carbohydrate, 23 g sugar, 7 g total fat, 17% calories from fat, 11 g fiber, 231 mg sodium

Southwest Lentil Mac
Serves 4

Mac n' Cheeze Sauce:

1 cup plain soymilk or almond milk
⅓ cup nutritional yeast
2 tablespoons cornstarch (or flour)
1 teaspoon onion powder
1 teaspoon garlic powder
½ teaspoon smoked paprika
⅛ teaspoon ground turmeric
1–2 tablespoons miso paste

2 cups uncooked elbow macaroni
1 cup cooked lentils
1 (8-ounce) jar salsa
1 (4-ounce) can diced green chiles
½ avocado, diced
Sliced green onions (optional)

For the Mac n' Cheeze Sauce: Whisk all ingredients except miso paste in a saucepan. Bring just to a boil over medium heat. Reduce to low and cook, stirring frequently, until slightly thickened. Turn off heat and stir in miso.

Meanwhile, in another pot, cook pasta as directed on package. Drain and rinse under cool water. Set aside.

Pulse lentils in a food processor to process into crumbles (optional). Fold cooked pasta into Mac n' Cheeze Sauce, stirring to coat completely. Fold in lentils, salsa, green chiles, avocado, and some green onions (if using). Garnish with additional green onions.

Per serving (¼ of recipe): 487 calories, 25 g protein, 85 g carbohydrate, 7 g sugar, 6 g total fat, 11% calories from fat, 12 g fiber, 679 mg sodium

Shanghai Noodles
Serves 2

2 sliced portabella mushroom caps, stems removed
1 tablespoon rice vinegar
1 tablespoon hoisin sauce
8 ounces thin dry Asian noodles (or linguine)
8 ounces frozen stir-fry vegetables
½ cup low-sodium vegetable broth (or water)
2 green onions, sliced, divided into white and green parts
1 tablespoon minced fresh ginger (about ½-inch piece)
1–2 minced garlic cloves
1–2 tablespoons Bragg Liquid Aminos
2 cups shredded cabbage
Sriracha

Toss mushrooms with vinegar and hoisin in a bowl. Let stand 10 minutes to marinate.

Cook noodles as directed on package, adding stir-fry vegetables 1–2 minutes before noodles are done cooking. Reserve ¼ cup of the cooking water. Drain and rinse noodles and vegetables under cool water. Set aside.

Meanwhile, pour ¼ cup broth (or water) into a large skillet over medium-high heat. Add white parts of green onions, ginger, and garlic, and cook until the onion is translucent. Add Bragg Liquid Aminos and stir to coat, scraping skillet with spatula to release any browned bits. Add mushrooms (with marinade) and ¼ cup broth (or water). Cook until mushrooms reach desired tenderness, stirring occasionally. Turn off heat and add shredded cabbage, stirring a few times so the cabbage softens. Stir in cooked noodles and vegetables, plus 1–2 tablespoons of the reserved cooking water to help loosen the sauce. Taste, adding more Bragg Liquid Aminos or hoisin as desired. Garnish with reserved green onions and a drizzle of sriracha.

Note: You can substitute low-sodium soy sauce for the Bragg Liquid Aminos.

Per serving (½ of recipe): 526 calories, 20 g protein, 109 g carbohydrate, 9 g sugar, 2 g total fat, 3% calories from fat, 15 g fiber, 586 mg sodium

Butternut Pasta

Serves 2

3–4 cups diced butternut squash (about 1 medium squash)
1 cup sliced mushrooms
1½ cups uncooked pasta
4 chopped dry-pack sun-dried tomatoes
1 cup canned low-sodium chickpeas or navy beans, drained and
    rinsed
2 cups fresh arugula (or spinach)
¼ cup fat-free balsamic vinaigrette
Vegan parmesan (optional)

Preheat oven to 375°F and line a baking sheet with parchment paper. Place squash cubes in a single layer in the pan. Roast for 20–30 minutes, or until fork-tender and browning at the edges, turning halfway. Add mushrooms to the pan halfway through roasting the squash.

Cook pasta as directed on package, adding sun-dried tomatoes and chickpeas to the water with the pasta. Drain and transfer to a bowl. Immediately add roasted squash and mushrooms and toss. Serve over (or mix in) arugula (or spinach) and drizzle with balsamic vinaigrette. Sprinkle with vegan parmesan, if desired.

Note: For homemade vegan parmesan recipe, see Vegan Parmesan, page 292.

Per serving (½ of recipe): 642 calories, 25 g protein, 126 g carbohydrate, 13 g sugar, 6 g total fat, 7% calories from fat, 17 g fiber, 351 mg sodium

Sides

Mexican Quinoa
Serves 1

½ cup uncooked quinoa (or brown rice)
½ cup frozen corn
½ cup fresh salsa
Cilantro (optional)

Cook quinoa or rice as directed on package. Add corn during last 5 minutes of cooking.

If your salsa is very chunky, puree half (optional). Combine warm quinoa with salsa, stirring to mix well. Fold in cilantro if using.

To make a meal out of this dish, stir in black beans, avocado, red onion, and hot sauce.

Per serving: 394 calories, 15 g protein, 74 g carbohydrate, 10 g sugar, 6 g total fat, 13% calories from fat, 9 g fiber, 410 mg sodium

Jerk Marinade
Makes ¼ cup

1½–2 tablespoons fresh lime juice
1 tablespoon low-sodium soy sauce
1 teaspoon pure maple syrup
1 teaspoon tomato paste
½ teaspoon dried thyme
½ teaspoon ground allspice
½ teaspoon ground cinnamon
½ teaspoon ground ginger
½ teaspoon garlic powder
Pinch ground cloves
Pinch red pepper flakes (optional)

Whisk all ingredients in a small bowl.

Usage Tip: Marinate 2 sliced portabella mushroom caps, stems removed, or 1 can chickpeas (drain and rinse first) in marinade for 24 hours (or longer). Cook in marinade until marinade has evaporated.

Serve over rice, as a burrito filling in tortillas, as tacos with corn tortillas, tomatoes, and guacamole, or on a salad with mixed greens, red onion, and pineapple.

Per tablespoon: 13 calories, 0.5 g protein, 3 g carbohydrate, 1 g sugar, 0.1 g total fat, 5% calories from fat, 0.5 g fiber, 145 mg sodium

Brazilian Rice
Serves 4

1 cup diced carrot
1 cup diced mango
1 (15-ounce) can low-sodium chickpeas (or pinto beans), drained and rinsed
1 cup cooked brown rice
1 (4-ounce) can diced green chiles (optional)
1 teaspoon paprika
½ teaspoon garlic powder
Pinch ground turmeric, for color (optional)
1–2 tablespoons fresh lime juice
1–2 tablespoons chopped cilantro (optional)
1 tablespoon low-sodium soy sauce or teriyaki sauce (optional)
1–2 teaspoons toasted sesame seeds

Pour ¼ cup water (or broth) into a large skillet over medium-high heat. Add carrot and cook until just fork-tender but still crisp. (If using frozen mango, add with raw carrots.) Add chickpeas, cooked rice, mango (if not using frozen), green chiles, paprika, garlic powder, and turmeric (if using), stirring to combine. Cook 1–2 minutes, or until heated through, adding a splash of water (or broth) as needed to prevent it from sticking. Turn off heat and stir in lime juice. Add cilantro and soy sauce, if desired. Sprinkle with sesame seeds before serving.

For a heartier dish, stir in crushed nuts (any kind) or peas.

Per serving (¼ of recipe): 199 calories, 7 g protein, 38 g carbohydrate, 10 g sugar, 3 g total fat, 13% calories from fat, 7 g fiber, 29 mg sodium

Basic Polenta

Serves 6

1½ cups polenta
1 cup plain soymilk or almond milk

Bring 3 cups water (or broth) to a boil in a medium saucepan. Reduce to low and slowly whisk in polenta. Whisk constantly to avoid clumps. Continue cooking and stirring until polenta is bubbly. Stir in soymilk or almond milk as needed for a creamy consistency. Season with salt and pepper to taste. You can also add nutritional yeast and/or Italian seasoning to taste. Add more milk just before serving for a creamier polenta.

Note: Add fresh lime juice from 1–2 limes for a variation.

Per serving (⅙ of recipe): 163 calories, 4 g protein, 33 g carbohydrate, 2 g sugar, 1 g total fat, 7% calories from fat, 2 g fiber, 178 mg sodium

Chipotle Slaw

Serves 3

¼–1 chipotle chile in adobo sauce (optional)
½ cup red wine vinegar
⅓ cup agave nectar (or maple syrup)
1–2 tablespoons adobo sauce (from the can of chipotle chiles)
1 teaspoon dried oregano
1 teaspoon garlic powder
1 (10–14 ounce) bag shredded cabbage or coleslaw mix

Combine all ingredients, except cabbage, in a blender and blend until smooth. Toss with cabbage just before serving.

For a meal, serve with avocado (or guacamole) plus black beans or refried beans over cooked rice, or in corn tortillas for tacos.

Note: Look for "chipotle chiles in adobo sauce," a canned item, in the international section of a grocery store.

Per serving (⅓ of recipe): 156 calories, 2 g protein, 36 g carbohydrate, 30 g sugar, 0.3 g total fat, 2% calories from fat, 2 g fiber, 68 mg sodium

Citrus Sweet Potatoes
Serves 5

2 cups diced sweet potatoes
1–2 teaspoons low-sodium soy sauce (optional)
1 cup diced red bell pepper
½ cup fresh mint, chopped
¼ cup sliced green onion or diced red onion
Juice of 1 orange
Juice of ½ small lemon
½ cup diced, peeled orange

Preheat oven to 400°F and line a baking sheet with parchment paper. Place sweet potatoes in a single layer in pan and roast 20–30 minutes, until fork-tender. Season with soy sauce, if desired, and add salt and pepper to taste. Toss sweet potatoes with bell pepper, mint, green onion, orange juice, and lemon juice. Garnish with diced oranges and serve. (Can be served warm or cold.)

For a stronger citrus flavor, you can add orange zest, but don't add too much or it will taste bitter.

Note: Omit mint and orange juice and add 1–2 teaspoons Bragg Liquid Aminos or low-sodium soy sauce for a variation.

Per serving (⅕ of recipe): 67 calories, 2 g protein, 15 g carbohydrate, 7 g sugar, 0.3 g total fat, 4% calories from fat, 3 g fiber, 167 mg sodium

Bayou Quinoa

Serves 4

½ cup uncooked quinoa
1 cup low-sodium vegetable broth
½ cup diced onion
½ cup diced celery
1 teaspoon salt-free Cajun seasoning
1 (14-ounce) can fire-roasted diced tomatoes (undrained)
1 (15-ounce) can low-sodium kidney beans, drained and rinsed
Sugar or ketchup (optional)
Louisiana-style hot sauce

Combine quinoa and 1 cup broth in a small saucepan. Cover and bring to a boil. Reduce heat to low and cook until quinoa is fluffy, about 15 minutes. Season with salt, pepper, and a few pinches of Cajun seasoning to taste, if desired. Set aside.

Meanwhile, pour ¼ cup water (or broth) into a large skillet over medium-high heat. Add onion and celery, and cook until onions are translucent. Stir in Cajun seasoning to coat. Cook 1 minute or until liquid evaporates. Turn off heat.

Pour half of the diced tomatoes (with juices) into a blender and puree until smooth or mostly smooth. Pour into skillet. Stir in remaining tomatoes (with juices) and beans. Cover and cook over low heat until heated through. Taste—if it's acidic, add sugar or ketchup. Season with salt, pepper, and more Cajun seasoning to taste. Stir in or serve over cooked quinoa with a drizzle of Louisiana-style hot sauce.

Per serving (¼ of recipe): 200 calories, 10 g protein, 36 g carbohydrate, 8 g sugar, 2 g total fat, 11% calories from fat, 8 g fiber, 506 mg sodium

Curried Rice

Serves 5

1 cup uncooked brown rice
2½ cups low-sodium vegetable broth
1½ teaspoons mild curry powder
½ teaspoon onion powder
¼ teaspoon chili powder
Pinch ground cumin
Pinch paprika
Pinch ground cinnamon
Pinch ground turmeric
2 tablespoons tomato sauce (or ketchup)

In a medium saucepan, combine rice with broth and all spices. Cover and bring to a boil. Reduce heat to low and simmer 40–45 minutes, or until rice is fully cooked. While still hot, stir in tomato sauce. Season with salt and pepper to taste.

To make a meal out of this dish, stir in steamed kale and carrots (or other mixed vegetables), plus raw cashews and peas (or kidney beans).

Per serving (⅕ of recipe): 164 calories, 4 g protein, 34 g carbohydrate, 1 g sugar, 1 g total fat, 7% calories from fat, 3 g fiber, 253 mg sodium

Perfect Brown Rice

Makes 3 cups

1 cup dry short-grain brown rice
3 cups water

Place the rice in a saucepan, rinse with water, then drain away the water. Place the pan on high heat and stir the rice until dry, about 2 minutes. Add the 3 cups water. Bring to a boil, then simmer until the rice is thoroughly cooked but still retains just a hint of crunchiness— about 40 minutes. Drain off the remaining water. Do not cook the rice until all the water is absorbed. Top with soy sauce, sesame seeds, cooked vegetables, beans, or lentils, if desired.

Per ½ cup serving: 115 calories, 2.7 g protein, 24 g carbohydrate, 0.4 g sugar, 1 g total fat, 7% calories from fat, 3 g fiber, 5 mg sodium

(Recipe by Neal Barnard)

Tofu Mayo

Makes 2 cups

1 package (12.3 ounces) extra-firm Mori-Nu tofu
2 tablespoons Dijon mustard
2 teaspoons white vinegar
⅛ teaspoon lemon juice
⅛ teaspoon agave nectar

In a blender or small food processor, blend tofu with Dijon mustard and vinegar until smooth and creamy. Add a few drops each of lemon juice and agave nectar and blend again. Season to taste with more lemon juice, agave, or mustard. Refrigerate until ready to use.

Note: In a pinch, blend tofu with juice from 1 lemon.

Per tablespoon: 8 calories, 1 g protein, 0.5 g carbohydrate, 0 g sugar, 0.5 g total fat, 40% calories from fat, 0 g fiber, 26 mg sodium

Vegan Parmesan

Makes 1 cup

1 cup cashews
½ cup nutritional yeast
Garlic powder and/or onion powder (optional)

Place cashews and nutritional yeast in a blender. Add garlic powder and/or onion powder, if desired, plus a pinch of salt, if desired. Process until a smooth powder is formed. Store in an airtight container in the fridge for a week.

Note: Use whatever nuts you have on hand, such as raw almonds, raw cashews, or raw Brazil nuts. It's also really good with ½ cup raw walnuts and ½ cup raw sunflower seeds or sesame seeds. Any combination of nuts and seeds will taste great. You can also start with less nutritional yeast (3–4 tablespoons), adding more to taste.

Per tablespoon: 63 calories, 4 g protein, 5 g carbohydrate, 0.5 g sugar, 4 g total fat, 51% calories from fat, 1 g fiber, 3 mg sodium

Desserts

Chocolate Cupcakes
Makes 12

1¼ cups flour
¼ cup unsweetened cocoa powder
1¼ teaspoons baking powder
1 teaspoon ground cinnamon
¾ teaspoon baking soda
½ teaspoon salt
1 ripe banana
½ cup applesauce
¼ cup brown sugar
¼ cup chocolate soymilk or almond milk
1 teaspoon vanilla extract
1 cup shredded zucchini
Chocolate Avocado Frosting (see below) or strawberry jam

Preheat oven to 350°F and line a muffin pan with paper liners or silicone cups, or use a nonstick pan. In a large bowl, whisk together flour, cocoa, baking powder, cinnamon, baking soda, and salt. In another bowl, mash banana with applesauce and sugar. Add milk, vanilla, and zucchini, and stir until evenly combined. Add dry ingredients in 3–4 batches and stir until just combined. Spoon batter evenly into muffin cups.

Bake 18–25 minutes, or until an inserted toothpick comes out clean. Cool completely. Spread with Chocolate Avocado Frosting or strawberry jam.

Note: You can stir ¼–½ cup chocolate chips into the batter.

Chocolate Avocado Frosting: Prepare just before spreading on cupcakes. Combine 1 ripe avocado with 3–4 tablespoons each unsweetened cocoa powder and pure maple syrup in a food processor. Process until completely smooth, adding a tiny bit of soymilk or almond milk if needed for consistency. Taste, adding more cocoa or maple syrup to taste, plus a pinch of salt, if desired. You can also add a few dashes of cinnamon or a few drops of vanilla extract.

(*continued*)

Per serving (¹⁄₁₂ of recipe): 126 calories, 3 g protein, 26 g carbohydrate, 11 g sugar, 2 g total fat, 16% calories from fat, 3 g fiber, 235 mg sodium

Fantastic Fruit Salsa with Cinnamon Sugar Chips
Serves 2

Cinnamon Sugar Chips:

3 (6-inch) corn tortillas (or 1 pita)
1–3 teaspoons cinnamon sugar

Fruit Salsa:

2 cups diced fruit, such as peaches, pineapple, watermelon, mango, strawberries, or a combination
¼ cup diced red onion
Juice of 1 lime
Minced jalapeño (optional)
Fresh cilantro (optional)

For the Cinnamon Sugar Chips: Preheat oven to 375°F and line a baking sheet with parchment paper. Cut tortillas or pita into triangles and place in a single layer on pan. Sprinkle with cinnamon sugar. Bake 5–10 minutes, or until chips are crisp.

For the Fruit Salsa: Mix all ingredients in a bowl until well combined. Serve with Cinnamon Sugar Chips.

Per serving (½ of recipe): 181 calories, 4 g protein, 42 g carbohydrate, 22 g sugar, 1 g total fat, 7% calories from fat, 5 g fiber, 19 mg sodium

Snickerdoodle Hummus
Serves 6

2 dates (or 5 tablespoons raisins)
1 (15-ounce) can low-sodium chickpeas, drained and rinsed
2 tablespoons peanut butter
2 tablespoons pure maple syrup
3 tablespoons plain soymilk or almond milk
2½ teaspoons vanilla extract
1½ teaspoons ground cinnamon

Soak dates in 1 cup hot water for 10 minutes, or until they plump. Drain, reserving soaking water.

In a food processor, combine chickpeas, soaked dates, peanut butter, maple syrup, milk, vanilla, and cinnamon. Process until smooth, adding more milk or soaking water as needed for desired consistency. (Soaking water will make it sweeter.) Taste, adding more peanut butter, maple syrup, or cinnamon as desired. This recipe is delicious served with apple slices.

Per serving (⅙ of recipe): 129 calories, 5 g protein, 19 g carbohydrate, 8 g sugar, 4 g total fat, 27% calories from fat, 4 g fiber, 31 mg sodium

Brownie Batter Hummus
Serves 6

1 (15-ounce) can low-sodium chickpeas, drained and rinsed
¼ cup plain soymilk or almond milk
¼ cup unsweetened cocoa powder
3 tablespoons agave nectar (or maple syrup)
1 teaspoon vanilla extract
Pinch salt
¼ cup vegan chocolate chips, melted (optional)

In a food processor, combine chickpeas, milk, cocoa powder, agave nectar (or maple syrup), vanilla, and salt. Process until smooth. Add more milk as needed for desired consistency. Transfer hummus to a serving bowl. Drizzle with melted chocolate chips and stir into hummus, if desired. Serve dip with crackers or apple slices.

Variations:

Mexican Chocolate Brownie Batter Hummus: Add ½ teaspoon ground cinnamon and a pinch of cayenne pepper with the vanilla.

Chocolate Mint Brownie Hummus: Substitute ½ teaspoon mint extract for the vanilla. Serve dip with crackers, rice cakes, pretzels, or pita chips.

Per serving (⅙ of recipe): 110 calories, 4 g protein, 21 g carbohydrate, 9 g sugar, 2 g total fat, 15% calories from fat, 4 g fiber, 58 mg sodium

Apple Pie Nachos

Serves 2

Cinnamon Sugar Chips:

3 (6-inch) corn tortillas (or 1 pita)
1–3 teaspoons cinnamon sugar

1 diced apple
½ teaspoon ground cinnamon
2 tablespoons agave nectar (or maple syrup)
5–6 ounces vegan yogurt (plain or vanilla)
Cinnamon, to taste
2 tablespoons crushed pecans (or walnuts)

For the Cinnamon Sugar Chips: Preheat oven to 375°F and line a baking sheet with parchment paper. Cut tortillas or pita into triangles and place in a single layer on pan. Sprinkle with cinnamon sugar. Bake 5–10 minutes, or until chips are crisp. Set aside.

Pour ¼ cup water into a skillet over low heat. Add apple and cinnamon. Cover and cook for about 1 hour or until diced apple is very soft and starting to break down. (You want it to be like a thick jam or compote.)

Place Cinnamon Sugar Chips on a platter and top with cooked apples. Drizzle with sweetener and top with a dollop of yogurt and a sprinkling of cinnamon (or serve yogurt on the side). Garnish with pecans (or walnuts).

Per serving (½ of recipe): 295 calories, 5 g protein, 55 g carbohydrate, 30 g sugar, 7 g total fat, 21% calories from fat, 6 g fiber, 27 mg sodium

Acknowledgments

Let me say a huge "Thank you!" to the many people who made this book possible. First, thank you to the many research volunteers who have participated in our studies. I am very grateful for all your early mornings, late evenings, and seemingly endless tests that have brought the power of nutrition to light. Thank you to the staff of the Physicians Committee for Responsible Medicine for making this research possible and for making sure that the world knows about it.

A special thank you to Lindsay S. Nixon, who lent her culinary expertise to this book and provided menus and recipes that are as delicious as they are healthful, and to Amber Green, RD, who provided expert nutrient analyses for each of them. Mandy and Jamie Gleason, Katie Fletcher, Reina Pohl, Nora Burgess, Esther Haugabrooks, Tawny Locke, Betsy Wason, Dania DePas, Elizabeth Baker, Namita Money, Claudia Elias, Meghan Jardine, and Andrea Cimino tested the recipes, and Natalie Hardcastle kept the process organized every step of the way. Natalie Hardcastle and Rosendo Flores also lent their considerable research skills. Thank you all!

Thank you to Cael Croft for providing expert illustrations and to Hana Kahleova, MD, PhD; Caroline Trapp, DNP; Kristi Funk, MD; Christie Mitchell Cobb, MD; Yoko Yokoyama, PhD; Mark Sklar, MD; Ali Haessler, MD; Ron Burmeister, MD; Kristine Slatkavitz,

297

FNP-BC; Bryan Cressey; Bonnie MacLeod; Erica Nielsen; Ashley Waddell; Emilie Rembert; Andrea Cimino; Mark Kennedy, Esq; Katie Fletcher; Willy Yonas; and Kenzie Phillips for sharing helpful insights and comments on the manuscript. Thank you to Dania DePas, Laura Anderson, and the PCRM Communications team, and to Linda Duggins for getting the word out so effectively. Thank you to Emily Rosman for keeping everything on track in the editorial process.

And an extra big thank you to Brian DeFiore, my literary agent, and to my editor, Leah Miller, for your support of this book.

I am especially grateful to Elsa, Katherine, Robin, Anna, Mary-Ann, Lindsay, Lee, Alison, Ray, Marie, Ann, Bob, Guy, Nancy, Mike, Wendy, Nina, Randa, Joy, Kim, April, Jeanne, and Sharmila for sharing their personal experiences. You will inspire many others.

References

Chapter 1: Foods for Fertility

1. Rich-Edwards JW, Goldman MB, Willett WC, et al. Adolescent body mass index and infertility caused by ovulatory disorder. *Am J Obstet Gynecol* 1994;171(1):171-177.

2. McGill CR, Fulgoni VL III, Devareddy L. Ten-year trends in fiber and whole grain intakes and food sources for the United States population: National Health and Nutrition Examination Survey 2001–2010. *Nutrients.* 2015;7(2):1119-1130.

3. Rose DP, Goldman M, Connolly JM, Strong LE. High-fiber diet reduces serum estrogen concentrations in premenopausal women. *Am J Clin Nutr.* 1991;54:520-525.

4. Goldin BR, Woods MN, Spiegelman DL, et al. The effect of dietary fat and fiber on serum estrogen concentrations in premenopausal women under controlled dietary conditions. *Cancer.* 1994;74(3 Suppl):1125-1131.

5. Bagga D, Ashley JM, Geffrey SP, et al. Effects of a very low fat, high fiber diet on serum hormones and menstrual function. Implications for breast cancer prevention. *Cancer.* 1995;76(12):2491-2496.

6. Carruba G, Granata OM, Pala V, et al. A traditional Mediterranean diet decreases endogenous estrogens in healthy postmenopausal women. *Nutr Cancer.* 2006;56(2):253-259.

7. Cramer DW, Xu H, Sahi T. Adult hypolactasia, milk consumption, and age-specific fertility. *Am J Epidemiol.* 1994;139(3):282-289.

8. Gross KC, Acosta PB. Fruits and vegetables are a source of galactose: implications in planning the diets of patients with galactosaemia. *J Inherit Metab Dis.* 1991;14(2):253-258.

9. Afeiche M, Williams PL, Mendiola J, et al. Dairy food intake in relation to semen quality and reproductive hormone levels among physically active young men. *Human Reproduction.* 2013;28(8):2265-2275.

10. Afeiche MC, Bridges ND, Williams PL, et al. Dairy intake and semen quality among men attending a fertility clinic. *Fertil Steril.* 2014;101(5):1280-1287.

11. Mendiola J, Torres-Cantero AM, Moreno-Grau JM, et al. Food intake and its relationship with semen quality: a case-control study. *Fertil Steril.* 2009;91:812-818.

12. Giahi L, Mohammadmoradi S, Javidan A, Sadeghi MR. Nutritional modifications in male infertility: a systematic review covering 2 decades. *Nutr Rev.* 2016;74(2):118-130.

13. Rich-Edwards JW, Spiegelman D, Garland M, et al. Physical activity, body mass index, and ovulatory disorder infertility. *Epidemiology.* 2002;13:184-190.

14. Shangold MM, Levine HS. The effect of marathon training upon menstrual function. *Am J Obstet Gynecol.* 1982;143(8):862-869.

15. American College of Obstetrics and Gynecology. Morning sickness: Nausea and vomiting of pregnancy. https://www.acog.org/Patients/FAQs/Morning-Sickness-Nausea-and-Vomiting-of-Pregnancy?IsMobileSet=false. Accessed April 4, 2019.

16. Hook EB. Changes in tobacco smoking and ingestion of alcohol and caffeinated beverages during early pregnancy: are these consequences, in part, of feto-protective mechanisms diminishing maternal exposure to embryotoxins? In: Kelly S, Hook EB, Janerich DT, Porter IH, eds. *Birth Defects: Risks and Consequences.* New York: Academic Press; 1976.

17. Cardwell MS. Pregnancy sickness: a biopsychological perspective. *Obstet Gynecol Surv.* 2012;67(10):645-652.

18. Fessler DM. Reproductive immunosuppression and diet. An evolutionary perspective on pregnancy sickness and meat consumption. *Curr Anthropol.* 2002;43(1):19-61.

19. Minturn L, Weiher AW. The influence of diet on morning sickness: a cross-cultural study. *Med Anthropol.* 1984;8(1):71-75.

20. Fessler DM. Reproductive immunosuppression and diet. An evolutionary perspective on pregnancy sickness and meat consumption. *Curr Anthropol.* 2002;43(1):19-61.

21. Signorello LB, Harlow BL, Wang S, Erick MA. Saturated fat intake and the risk of severe hyperemesis gravidarum. *Epidemiology.* 1998;9(6):636-640.

Chapter 2: Curing Cramps and Premenstrual Syndrome

1. Barnard ND, Scialli AR, Hurlock D, Bertron P. Diet and sex-hormone binding globulin, dysmenorrhea, and premenstrual symptoms. *Obstet Gynecol.* 2000;95:245-250.

2. Marshall LM, Spiegelman D, Manson JE, et al. Risk of uterine leiomyomata among premenopausal women in relation to body size and cigarette smoking. *Epidemiology.* 1998;9:511-517.

3. Wise LA, Palmer JR, Spiegelman D, et al. Influence of body size and body fat distribution on risk of uterine leiomyomata in U.S. black women. *Epidemiology*. 2005;16(3):346-354.

4. Lee JE, Song S, Cho E, et al. Weight change and risk of uterine leiomyomas: Korea Nurses' Health Study. *Curr Med Res Opin*. 2018;9:1-7.

5. Parazzini F, Di Martino M, Candiani M, Viganò P. Dietary components and uterine leiomyomas: a review of published data. *Nutrition and Cancer*. 2015;67(4):569-579.

Chapter 3: Tackling Cancer for Women

1. Yager JD, Davidson NE. Estrogen carcinogenesis in breast cancer. *N Engl J Med*. 2006;354:270-282.

2. Endogenous Hormones and Breast Cancer Collaborative Group. Endogenous sex hormones and breast cancer in postmenopausal women: reanalysis of nine prospective studies. *J Natl Cancer Inst*. 2002;94:606-616.

3. Wynder EL, Kajitani T, Kuno J, Lucas JC Jr, DePalo A, Farrow J. A comparison of survival rates between American and Japanese patients with breast cancer. *Surg Gynec Obstet*. 1963;117:196-200.

4. Saika K, Sobue T. Epidemiology of breast cancer in Japan and the US. *JMAJ*. 2009;52(1):39-44.

5. Shin S, Saito E, Inoue M, et al. Dietary pattern and breast cancer risk in Japanese women: the Japan Public Health Center-based Prospective Study (JPHC Study). *Br J Nutr*. 2016;115:1769-1779.

6. Rose DP, Goldman M, Connolly JM, Strong LE. High-fiber diet reduces serum estrogen concentrations in premenopausal women. *Am J Clin Nutr*. 1991;54:520-525.

7. Goldin BR, Woods MN, Spiegelman DL, et al. The effect of dietary fat and fiber on serum estrogen concentrations in premenopausal women under controlled dietary conditions. *Cancer*. 1994;74(3 Suppl):1125-1131.

8. Bagga D, Ashley JM, Geffrey SP, et al. Effects of a very low fat, high fiber diet on serum hormones and menstrual function. Implications for breast cancer prevention. *Cancer*. 1995;76(12):2491-2496.

9. Nagata C, Nagao Y, Shibuya C, Kashiki Y, Shimizu H. Fat intake is associated with serum estrogen and androgen concentrations in postmenopausal Japanese women. *J Nutr*. 2005;135:2862-2865.

10. Gregorio DI, Emrich LJ, Graham S, Marshall JR, Nemoto T. Dietary fat consumption and survival among women with breast cancer. *J Natl Cancer Inst*. 1985;75:37-41.

11. Chlebowski RT, Blackburn GL, Thomson CA, et al. Dietary fat reduction and breast cancer outcome: interim efficacy results from the Women's Intervention Nutrition Study. *J Natl Cancer Inst*. 2006;98:1767-1776.

12. Nomura A, Le Marchand L, Kolonel LN, Hankin JH. The effect of dietary fat on breast cancer survival among Caucasian and Japanese women in Hawaii. *Breast Cancer Res Treat.* 1991;18:S135-S141.

13. Zhang S, Folsom AR, Sellers TA, Kushi LH, Potter JD. Better breast cancer survival for postmenopausal women who are less overweight and eat less fat. *Cancer.* 1995;76:275-283.

14. Thomas HV, Davey GK, Key TJ. Oestradiol and sex hormone-binding globulin in premenopausal and post-menopausal meat-eaters, vegetarians and vegans. *Br J Cancer.* 1999;80(9):1470-1475.

15. Karelis AD, Fex A, Filion ME, Adlercreutz H, Aubertin-Leheudre M. Comparison of sex hormonal and metabolic profiles between omnivores and vegetarians in pre- and post-menopausal women. *Br J Nutr.* 2010;104(2):222-226.

16. Barnard ND, Scialli AR, Hurlock D, Bertron P. Diet and sex-hormone binding globulin, dysmenorrhea, and premenstrual symptoms. *Obstet Gynecol.* 2000;95:245-250.

17. Brinkman MT, Baglietto L, Krishnan K, et al. Consumption of animal products, their nutrient components and postmenopausal circulating steroid hormone concentrations. *Eur J Clin Nutr.* 2010;64(2):176-183.

18. Kroenke CH, Kwan ML, Sweeney C, Castillo A, Caan BJ. High- and low-fat dairy intake, recurrence, and mortality after breast cancer diagnosis. *J Natl Cancer Inst.* 2013;105:616-623.

19. Dong JY, He K, Wang P, Qin LQ. Dietary fiber intake and risk of breast cancer: a meta-analysis of prospective cohort studies. *Am J Clin Nutr.* 2011;94:900-905.

20. Holmes MD, Liu S, Hankinson SE, Colditz GA, Hunter DJ, Willett WC. Dietary carbohydrates, fiber, and breast cancer risk. *Am J Epidemiol.* 2004;159:732-739.

21. Farvid MS, Stern MC, Norat T, et al. Consumption of red and processed meat and breast cancer incidence: a systematic review and meta-analysis of prospective studies. *Int J Cancer.* 2018 Sep 5. doi: 10.1002/ijc.31848. [Epub ahead of print]

22. McTiernan A, Wu L, Chen C, et al. Relation of BMI and physical activity to sex hormones in postmenopausal women. *Obesity.* 2006;14:1662-1677.

23. Chlebowski RT, Luo J, Anderson GL, et al. Weight loss and breast cancer incidence in postmenopausal women. *Cancer.* 2019;125(2):205-212.

24. Zhang X, Eliassen AH, Tamimi RM, et al. Adult body size and physical activity in relation to risk of breast cancer according to tumor androgen receptor status. *Cancer Epidemiol Biomarkers Prev.* 2015;24(6):962-968.

25. Suzuki R, Rylander-Rudqvist T, Ye W, Saji S, Wolk A. Body weight and postmenopausal breast cancer risk defined by estrogen and progesterone receptor status among Swedish women: a prospective cohort study. *Int. J. Cancer.* 2006;119:1683-1689.

26. Rock CL, Demark-Wahnefried W. Nutrition and survival after the diagnosis of breast cancer: a review of the evidence. *J Clin Oncol.* 2002;20:3302-3316.

27. Tao MH, Shu XO, Ruan ZX, Gao YT, Zheng W. Association of overweight with breast cancer survival. *Am J Epidemiol.* 2006;163:101-107.

28. Academy of Nutrition and Dietetics. Is it safe to take antioxidant supplements during chemotherapy and radiation therapy? April 2013. https://www.oncologynutrition.org/erfc/eating-well-when-unwell/antioxidant-supplements-safe-during-therapy. Accessed April 15, 2019.

29. Pierce JP, Faerber S, Wright FA, et al. A randomized trial of the effect of a plant-based dietary pattern on additional breast cancer events and survival: the Women's Healthy Eating and Living (WHEL) Study. *Contr Clin Trials.* 2002;23:728-756.

30. Rock CL, Flatt SW, Thomson CA, et al. Effects of a high-fiber, low-fat diet intervention on serum concentrations of reproductive steroid hormones in women with a history of breast cancer. *J Clin Oncol.* 2004;12:2379-2387.

31. Rock CL, Flatt SW, Natarajan L, et al. Plasma carotenoids and recurrence-free survival in women with a history of breast cancer. *J Clin Oncol.* 2005; 23:6631-6638.

32. Pierce JP, Stefanick ML, Flatt SW, et al. Greater survival after breast cancer in physically active women with high vegetable-fruit intake regardless of obesity. *J Clin Oncol.* 2007;25:2345-2351.

33. Pierce JP, Natarajan L, Caan BJ, et al. Influence of a diet very high in vegetables, fruit, and fiber and low in fat on prognosis following treatment for breast cancer: the Women's Healthy Eating and Living (WHEL) randomized trial. *JAMA.* 2007;298:289-298.

34. Fung TT, Chiuve SE, Willett WC, Hankinson SE, Hu FB, Holmes MD. Intake of specific fruits and vegetables in relation to risk of estrogen receptor-negative breast cancer among postmenopausal women. *Breast Cancer Res Treat.* 2013;138:925-930.

35. Xie Q, Chen ML, Qin Y, et al. Isoflavone consumption and risk of breast cancer: a dose-response meta-analysis of observational studies. *Asia Pac J Clin Nutr.* 2013;22(1):118-127.

36. Chen M, Rao Y, Zheng Y, et al. Association between soy isoflavone intake and breast cancer risk for pre- and post-menopausal women: a meta-analysis of epidemiological studies. *PLoS ONE.* 2014;9(2):e89288.

37. Nechuta SJ, Caan BJ, Chen WY, et al. Soy food intake after diagnosis of breast cancer and survival: an in-depth analysis of combined evidence from cohort studies of US and Chinese women. *Am J Clin Nutr.* 2012;96:123-132.

38. Chi F, Wu R, Zeng YC, Xing R, Liu Y, Xu ZG. Post-diagnosis soy food intake and breast cancer survival: a meta-analysis of cohort studies. *Asian Pac J Cancer Prev.* 2013;14(4):2407-2412.

39. World Cancer Research Fund/American Institute for Cancer Research. Continuous Update Project Expert Report 2018. Diet, Nutrition, Physical Activity and Breast Cancer. http://www.dietandcancerreport.org.

40. Singletary KW, Gapstur SM. Alcohol and breast cancer: review of epidemiologic and experimental evidence and potential mechanisms. *JAMA.* 2001 Nov 7;286(17):2143-2151.

41. Yamamoto M, Patel NA, Taggart J, Sridhar R, Cooper DR. A shift from normal to high glucose levels stimulates cell proliferation in drug sensitive MCF-7 human breast cancer cells but not in multidrug resistant MCF-7/ADR cells which overproduce PKC-β. *Int J Cancer.* 1999;83:98-106.

42. La Vecchia C, Giordano SH, Hortobagyi GN, Chabner B. Overweight, obesity, diabetes, and risk of breast cancer: interlocking pieces of the puzzle. *Oncologist.* 2011;16:726-729.

43. McTiernan A, Tworoger SS, Ulrich CM, et al. Effect of exercise on serum estrogen in postmenopausal women: a 12-month randomized clinical trial. *Cancer Res.* 2004a;64:2923-2928.

44. McTiernan A, Tworoger S, Rajan B, et al. Effect of exercise on serum androgens in postmenopausal women: a 12-month randomized clinical trial. *Cancer Epidemiol Biomarkers Prev.* 2004b;13:1099-1105.

45. World Cancer Research Fund/American Institute for Cancer Research. Continuous Update Project Expert Report 2018. Diet, Nutrition, Physical Activity and Breast Cancer. http://www.dietandcancerreport.org.

46. Mørch LS, Skovlund CW, Hannaford PC, Iversen L, Fielding S, Lidegaard Ø. Contemporary hormonal contraception and the risk of breast cancer. *N Engl J Med.* 2017;377(23):2228-2239.

47. Zolfaroli I, Tarín JJ, Cano A. Hormonal contraceptives and breast cancer: clinical data. *Eur J Obstet Gynecol Reprod Biol.* 2018;230:212-216.

48. Ibid.

49. Samson M, Porter N, Orekoya O, et al. Progestin and breast cancer risk: a systematic review. *Breast Cancer Res Treat.* 2016;155(1):3-12.

50. Prema K, Lakshmi BA, Babu S. Serum copper in long-term users of copper intrauterine devices. *Fertil Steril.* 1980;34(1):32-35.

51. De la Cruz D, Cruz A, Arteaga M, Castillo L, Tovalin H. Blood copper levels in Mexican users of the T380A IUD. *Contraception.* 2005;72:122-125.

52. Fahmy K, Ghoneim M, Eisa I, El-Gazar A, Afifi A. Serum and endometrial copper, zinc, iron and cobalt with inert and copper-containing IUCDs. *Contraception* 1993;47:483-490.

53. Imani S, Moghaddam-Banaem L, Roudbar-Mohammadi S, Asghari-Jafarabadi M. Changes in copper and zinc serum levels in women wearing a copper TCu-380A intrauterine device. *Eur J Contracept Reprod Health Care.* 2014;19(1):45-50.

54. Mathys ZK, White AR. Copper and Alzheimer's disease. *Adv Neurobiol.* 2017;18:199-216.

55. Sensi SL, Granzotto A, Siotto M, Squitti R. Copper and zinc dysregulation in Alzheimer's disease. *Trends Pharmacol Sci.* 2018;39(12):1049-1063.

56. Crandall CJ, Hovey KM, Andrews CA, et al. Breast cancer, endometrial cancer, and cardiovascular events in participants who used vaginal estrogen in the Women's Health Initiative Observational Study. *Menopause.* 2018;25(1):11-20.

57. Lukanova A, Lundin E, Micheli A, et al. Circulating levels of sex steroid hormones and risk of endometrial cancer in postmenopausal women. *Int J Cancer.* 2004;108:425-432.

58. Dougan MM, Hankinson SE, Vivo ID, Tworoger SS, Glynn RJ, Michels KB. Prospective study of body size throughout the life-course and the incidence of endometrial cancer among premenopausal and postmenopausal women. *Int J Cancer.* 2015;137(3):625-637.

59. World Cancer Research Fund/American Institute for Cancer Research. Continuous Update Project Expert Report 2018. Diet, Nutrition, Physical Activity and Endometrial Cancer. http:www.dietandcancerreport.org.

60. Rota M, Pasquali E, Scotti L, et al. Alcohol drinking and epithelial ovarian cancer risk. a systematic review and meta-analysis. *Gynecol Oncol.* 2012 Jun;125(3):758-763.

61. Jordan SJ, Whiteman DC, Purdie DM, Green AC, Webb PM. Does smoking increase risk of ovarian cancer? A systematic review. *Gynecol Oncol.* 2006 Dec;103(3):1122-1129.

62. Cramer DW. Lactase persistence and milk consumption as determinants of ovarian cancer risk. *Am J Epidemiol.* 1989;130(5):904-910.

63. Genkinger JM, Hunter DJ, Spiegelman D, et al. A pooled analysis of 12 cohort studies of dietary fat, cholesterol and egg intake and ovarian cancer. *Cancer Causes Control.* 2006;17(3):273-285.

64. Qin B, Moorman PG, Alberg AJ, et al. Dairy, calcium, vitamin D and ovarian cancer risk in African-American women. *Br J Cancer.* 2016;115:1122-1130. doi:10.1038/bjc.2016.289.

65. Larsson SC, Bergkvist L, Wolk A. Milk and lactose intakes and ovarian cancer risk in the Swedish Mammography Cohort. *Am J Clin Nutr.* 2004;80(5):1353-1357.

66. Larsson SC, Orsini N, Wolk A. Milk, milk products and lactose intake and ovarian cancer risk: a meta-analysis of epidemiological studies. *Int J Cancer.* 2006;118(2):431-441.

Chapter 4: Tackling Cancer for Men

1. World Cancer Research Fund/American Institute for Cancer Research. Continuous Update Project Expert Report 2018. Diet, Nutrition, Physical Activity and Prostate Cancer. Available at dietandcancerreport.org.

2. Ganmaa D, Li X, Wang J, Qin L, Wang P, Sato A. Incidence and mortality of testicular and prostatic cancers in relation to world dietary practices. *Int J Cancer.* 2002:98,262-267.

3. Monn MF, Tatem AJ, Tatem J, Cheng L. Prevalence and management of prostate cancer among East Asian men: current trends and future perspectives. *Urol Oncol.* 2016;34(2):58.e1-58.e9.

4. Chan JM, Stampfer MJ, Ma J, Gann PH, Gaziano JM, Giovannucci EL. Dairy products, calcium, and prostate cancer risk in the Physicians' Health Study. *Am J Clin Nutr.* 2001;74:549-554.

5. Giovannucci E, Rimm EB, Wolk A, et al. Calcium and fructose intake in relation to risk of prostate cancer. *Cancer Res.* 1998;58:442-447.

6. Lu W, Chen H, Niu Y, Wu H, Xia D, Wu Y. Dairy products intake and cancer mortality risk: a meta-analysis of 11 population-based cohort studies. *Nutr J.* 2016;15:91.

7. Heaney RP, McCarron DA, Dawson-Hughes B, et al. Dietary changes favorably affect bone remodeling in older adults. *J Am Diet Assoc.* 1999;99:1228-1233.

8. Chan JM, Stampfer MJ, Giovannucci E, et al. Plasma insulin-like growth factor-I and prostate cancer risk: a prospective study. *Science.* 1998;279:563-566.

9. Chavarro JE, Stampfer MJ, Li H, Campos H, Kurth T, Ma J. A prospective study of polyunsaturated fatty acid levels in blood and prostate cancer risk. *Cancer Epidemiol Biomarkers Prev.* 2007;16(7):1364-1370.

10. Brasky TM, Darke AK, Song X, et al. Plasma phospholipid fatty acids and prostate cancer risk in the SELECT Trial. *J Natl Cancer Inst.* 2013;105:1132-1141.

11. Crowe FL, Appleby PN, Travis RC, et al. Circulating fatty acids and prostate cancer risk: individual participant meta-analysis of prospective studies. *J Natl Cancer Inst.* 2014;106(9):dju240 doi:10.1093/jnci/dju240.

12. Giovannucci E, Rimm EB, Liu Y, Stampfer MJ, Willett WC. A prospective study of tomato products, lycopene, and prostate cancer risk. *J Natl Cancer Inst.* 2002;94(5):391-398.

13. Rowles JL 3rd, Ranard KM, Smith JW, An R, Erdman JW Jr. Increased dietary and circulating lycopene are associated with reduced prostate cancer risk: a systematic review and meta-analysis. *Prostate Cancer Prostatic Dis.* 2017;20(4):361-377.

14. Applegate CC, Rowles JL, Ranard KM, Jeon S, Erdman JW. Soy consumption and the risk of prostate cancer: an updated systematic review and meta-analysis. *Nutrients.* 2018;10(1). pii: E40. doi: 10.3390/nu10010040.

15. Robbins AS, Koppie TM, Gomez SL, Parikh-Patel A, Mills PK. Differences in prognostic factors and survival among white and Asian men with prostate cancer, California, 1995–2004. *Cancer.* 2007;110:1255-1263.

16. Ornish D, Weidner G, Fair WR, et al. Intensive lifestyle changes may affect the progression of prostate cancer. *J Urol.* 2005 Sep;174(3):1065-1070.

17. Frattaroli J, Weidner G, Dnistrian AM, et al. Clinical events in prostate cancer lifestyle trial: results from two years of follow-up. *Urology.* 2008;72(6):1319-1323.

18. Giannandrea F, Paoli D, Figà-Talamanca I, Lombardo F, Lenzi A, Gandini L. Effect of endogenous and exogenous hormones on testicular cancer: the epidemiological evidence. *Int J Dev Biol*. 2013;57:255-263.

19. Garner MJ, Birkett NJ, Johnson KC, et al. Dietary risk factors for testicular carcinoma. *Int J Cancer*. 2003 Oct 10;106(6):934-941.

Chapter 5: Reversing Polycystic Ovary Syndrome

1. Stein IF, Leventhal ML. Amenorrhea associated with bilateral polycystic ovaries. *Am J Obstet Gynecol*. 1935;29:181-191.

2. Alpañés M, Fernández-Durán E, Escobar-Morreale HF. Androgens and polycystic ovary syndrome. *Expert Rev Endocrinol Metab*. 2012;7(1):91-102.

3. American College of Obstetrics and Gynecology. ACOG Practice Bulletin. Polycystic ovary syndrome. *Obstet Gynecol*. 2018;131(6):e157-e171.

4. Clark JL, Taylor CG, Zahradka P. Rebelling against the (insulin) resistance: a review of the proposed insulin-sensitizing actions of soybeans, chickpeas, and their bioactive compounds. *Nutrients*. 2018;10(4). pii: E434. doi: 10.3390/nu10040434.

5. Jamilian M, Asemi Z. The effects of soy isoflavones on metabolic status of patients with polycystic ovary syndrome. *J Clin Endocrinol Metab*. 2016; 101(9):3386-3394.

6. Karamali M, Kashanian M, Alaeinasab S, Asemi Z. The effect of dietary soy intake on weight loss, glycaemic control, lipid profiles and biomarkers of inflammation and oxidative stress in women with polycystic ovary syndrome: a randomised clinical trial. *J Hum Nutr Diet*. 2018;31(4):533-543.

7. Botwood N, Hamilton-Fairley D, Kiddy D, Robinson S, Franks S. Sex hormone-binding globulin and female reproductive function. *J Steroid Biochem Mol Biol*. 1995;53(1-6):529-531.

8. Holt SHA, Brand Miller JC, Petocz P. An insulin index of foods: the insulin demand generated by 1000-kJ portions of common foods. *Am J Clin Nutr*. 1997;66:1264-1276.

Chapter 6: Tackling Menopause

1. Gail Sheehy. *The Silent Passage: Menopause*. New York: Random House; 1991:30.

2. Lock M. Menopause: lessons from anthropology. *Psychosomatic Med*. 1998;60:410-419.

3. Shea JL. Cross-cultural comparison of women's midlife symptom-reporting: a China study. *Cult Med Psychiatry*. 2006;30:331-362.

4. Melby MK, Lock M, Kaufert P. Culture and symptom reporting at menopause. *Hum Reprod Update*. 2005;11:495-512.

5. Beyenne Y, Martin MC. Menopausal experiences and bone density of Mayan women in Yucatan, Mexico. *Am J Human Biol.* 2001;13:505-511.

6. Melby MK. Vasomotor symptom prevalence and language of menopause in Japan. *Menopause.* 2005;12:250-257.

7. Smith DC, Prentice R, Thompson DJ, Herrmann WL. Association of exogenous estrogen and endometrial carcinoma. *N Engl J Med.* 1975;293(23):1164-1167.

8. Ziel HK, Finkle WD. Increased risk of endometrial carcinoma among users of conjugated estrogens. *N Engl J Med.* 1975;293(23):1167-1170.

9. American College of Obstetrics and Gynecology. Response to Women's Health Initiative Study Results by the American College of Obstetricians and Gynecologists, August 9, 2002. http://www.losolivos-obgyn.com/info/gynecology/menopause/acog_whi_2002.pdf. Accessed September 3, 2018.

10. Files JA, Ko MG, Pruthi S. Bioidentical hormone therapy. *Mayo Clin Proc.* 2011;86(7):673-680.

11. Roth JA, Etzioni R, Waters TM, et al. Economic return from the Women's Health Initiative estrogen plus progestin clinical trial: a modeling study. *Ann Intern Med.* 2014;160(9):594-602.

12. Power ML, Anderson BL, Schulkin J. Attitudes of obstetrician-gynecologists towards the evidence from the WHI HT trials remain generally skeptical. *Menopause.* 2009;16(3):500-508.

13. McIntosh J, Blalock SJ. Effects of media coverage of Women's Health Initiative study on attitudes and behavior of women receiving hormone replacement therapy. *Am J Health Syst Pharm.* 2005;62:69-74.

14. Chlebowski RT, Anderson GL, Gass M, et al. Estrogen plus progestin and breast cancer incidence and mortality in postmenopausal women. *JAMA.* 2010;304(15):1684-1692.

15. Manson JE, Aragaki AK, Rossouw JE. Menopausal hormone therapy and long-term all-cause and cause-specific mortality: the Women's Health Initiative randomized trials. *JAMA.* 2017;318(10):927-938.

16. US Preventive Services Task Force. Hormone therapy for the primary prevention of chronic conditions in postmenopausal women: US Preventive Services Task Force Recommendation Statement. *JAMA.* 2017;318(22):2224-2233.

17. Equine Ranching Advisory Board. Care and oversight of horses managed for the collection of pregnant mares' urine (PMU). 2014. http://www.naeric.org/assets/pdf/PMU-WhitePaper.pdf. Accessed September 13, 2018.

18. Files JA, Ko MG, Pruthi S. Bioidentical hormone therapy. *Mayo Clin Proc.* 2011;86(7):673-680.

19. Endogenous Hormones and Breast Cancer Collaborative Group. Endogenous sex hormones and breast cancer in postmenopausal women: reanalysis of nine prospective studies. *J Natl Cancer Inst.* 2002;94:606-616.

20. Thurston RC, Joffe H. Vasomotor symptoms and menopause: findings from the Study of Women's Health Across the Nation. *Obstet Gynecol Clin N Am.* 2011;38:489-501.

21. Nagata C, Shimizu H, Takami R, Hayashi M, Takeda N, Yasuda K. Hot flashes and other menopausal symptoms in relation to soy product intake in Japanese women. *Climacteric.* 1999;2:6-12.

22. Nagata C, Takatsuka N, Kawakami N, Shimuzu H. Soy product intake and hot flashes in Japanese women: results from a community-based prospective study. *Am J Epidemiol.* 2001;153:790-793.

23. Murkies AL, Lombard C, Strauss BJ, Wilcox G, Burger HG, Morton MS. Dietary flour supplementation decreases post-menopausal hot flashes: effect of soy and wheat. *Maturitas.* 2008;61(1-2):27-33.

24. Lewis JE, Nickell LA, Thompson LU, Szalai JP, Kiss A, Hilditch JR. A randomized controlled trial of the effect of dietary soy and flaxseed muffins on quality of life and hot flashes during menopause. *Menopause.* 2006;13(4):631-642.

25. Lethaby A, Marjoribanks J, Kronenberg F, Roberts H, Eden J, Brown J. Phytoestrogens for menopausal vasomotor symptoms. *Cochrane Database Syst Rev.* 2013;(12):CD001395. doi: 10.1002/14651858.CD001395.pub4.

26. Franco OH, Chowdhury R, Troup J, et al. Use of plant-based therapies and menopausal symptoms: a systematic review and meta-analysis. *JAMA.* 2016; 315(23):2554-2563.

27. Ibid.

28. Ghazanfarpour M, Sadeghi R, Roudsari RL, Khorsand I, Khadivzadeh T, Muoio B. Red clover for treatment of hot flashes and menopausal symptoms: a systematic review and meta-analysis. *J Obstet Gynaecol.* 2016;36:301-311.

29. Gartoulla P, Han, MM. Red clover extract for alleviating hot flashes in postmenopausal women: a meta-analysis. *Maturitas.* 2014;79:58-64.

30. Mehrpooya M, Rabiee S, Larki-Harchegani A, et al. A comparative study on the effect of "black cohosh" and "evening primrose oil" on menopausal hot flashes. *J Educ Health Promot.* 2018;7:36.

31. Crandall CJ, Hovey KM, Andrews CA, et al. Breast cancer, endometrial cancer, and cardiovascular events in participants who used vaginal estrogen in the Women's Health Initiative Observational Study. *Menopause.* 2018;25(1):11-20.

32. Meixel A, Yanchar E, Fugh-Berman A. Hypoactive sexual desire disorder: inventing a disease to sell low libido. *J Med Ethics.* 2015;41(10):859-862.

33. Hurst BS, Jones AI, Elliot M, Marshburn PB, Matthews ML. Absorption of vaginal estrogen cream during sexual intercourse: a prospective, randomized, controlled trial. *J Reprod Med.* 2008;53(1):29-32.

34. Franco OH, Chowdhury R, Troup J, et al. Use of plant-based therapies and menopausal symptoms: a systematic review and meta-analysis. *JAMA.* 2016;315(23):2554-2563.

35. Dizavandi FR, Ghazanfarpour M, Roozbeh N, Kargarfard L, Khadivzadeh T, Dashti S. An overview of the phytoestrogen effect on vaginal health and dyspareunia in peri- and post-menopausal women. *Post Reprod Health*. 2019;25(1): 11-20.

36. Najaf Najafi M, Ghazanfarpour M. Effect of phytoestrogens on sexual function in menopausal women: a systematic review and meta-analysis. *Climacteric*. 2018;21(5):437-445.

37. Freeman EW, Sammel MD, Lin H, Nelson DB. Associations of hormones and menopausal status with depressed mood in women with no history of depression. *Arch Gen Psychiatry*. 2006;63(4):375-382.

Chapter 7: Curing Erectile Dysfunction and Saving Your Life

1. Gowani Z, Uddin SMI, Mirlolouk M, et al. Vascular erectile dysfunction and subclinical cardiovascular disease. *Curr Sex Health Rep*. 2017;9(4):305-312.

2. Layton JB, Kim Y, Alexander GC, et al. Association between direct-to-consumer advertising and testosterone testing and initiation in the United States, 2009–2013. *JAMA*. 2017;317:1159-1166.

3. Bhasin S, Brito JP, Cunningham GR, et al. Testosterone therapy in men with hypogonadism: an endocrine society clinical practice guideline. *J Clin Endocrinol Metab*. 2018;103:1715-1744.

4. Layton JB, Li D, Meier CR, Sharpless JL, Stürmer T, Brookhart MA. Injection testosterone and adverse cardiovascular events: a case-crossover analysis. *Clin Endocrinol (Oxf)*. 2018;88(5):719-727.

5. Ornish D, Brown SE, Scherwitz LW, Billings JH, Armstrong WT, Ports TA. Can lifestyle changes reverse coronary heart disease? *Lancet*. 1990;336:129-133.

6. Esposito K, Giugliano F, Di Palo C, et al. Effect of lifestyle changes on erectile dysfunction in obese men: a randomized controlled trial. *JAMA*. 2004;291:2978-2984.

7. Chiavaroli L, Nishi SK, Khan TA, et al. Portfolio dietary pattern and cardiovascular disease: a systematic review and meta-analysis of controlled trials. *Prog Cardiovasc Dis*. 2018;61:43-53.

8. Derby CA, Mohr BA, Goldstein I, Feldman HA, Johannes CB, McKinlay JB. Modifiable risk factors and erectile dysfunction: can lifestyle changes modify risk? *Urology*. 2000;56(2):302-306.

9. RxList. Viagra. https://www.rxlist.com/viagra-drug.htm#description. Accessed August 22, 2018.

Chapter 8: Conquering Diabetes

1. Tonstad S, Butler T, Yan R, Fraser GE. Type of vegetarian diet, body weight, and prevalence of type 2 diabetes. *Diabetes Care*. 2009;32:791-796.

2. Barnard ND, Cohen J, Jenkins DJ, et al. A low-fat, vegan diet improves glycemic control and cardiovascular risk factors in a randomized clinical trial in individuals with type 2 diabetes. *Diabetes Care.* 2006;29:1777-1783.

3. Petersen KF, Dufour S, Befroy D, Garcia R, Shulman GI. Impaired mitochondrial activity in the insulin-resistant offspring of patients with type 2 diabetes. *N Engl J Med.* 2004;350:664-671.

4. Bunner AE, Wells CL, Gonzales J, Agarwal U, Bayat E, Barnard ND. A dietary intervention for chronic diabetic neuropathy pain: a randomized controlled pilot study. *Nutr Diabetes.* 2015;5:e158. doi: 10.1038/nutd.2015.8.

5. Karjalainen J, Martin JM, Knip M, et al. A bovine albumin peptide as a possible trigger of insulin-dependent diabetes mellitus. *N Engl J Med.* 1992;327(5):302-307.

Chapter 9: A Healthy Thyroid

1. Kratzsch J, Fiedler GM, Leichtle A, et al. New reference intervals for thyrotropin and thyroid hormones based on National Academy of Clinical Biochemistry criteria and regular ultrasonography of the thyroid. *Clin Chem.* 2005;51:1480-1486.

2. Feyrer J, Politi D, Weil DN. The cognitive effects of micronutrient deficiency: evidence from salt iodization in the United States. *J Eur Econ Assoc.* 2017; 15(2):355-387. doi:10.1093/jeea/jvw002.

3. Zava TT, Zava DT. Assessment of Japanese iodine intake based on seaweed consumption in Japan: a literature-based analysis. *Thyroid Res.* 2011;4:14. doi.org/10.1186/1756-6614-4-14.

4. Laurberg P, Bülow Pedersen I, Knudsen N, Ovesen L, Andersen S. Environmental iodine intake affects the type of nonmalignant thyroid disease. *Thyroid.* 2001;11:457-469.

5. Desailloud R, Hober D. Viruses and thyroiditis: an update. *Virol J.* 2009;6:5. doi:10.1186/1743-422X-6-5.

6. Matana A, Torlak V, Brdar D, et al. Dietary factors associated with plasma thyroid peroxidase and thyroglobulin antibodies. *Nutrients.* 2017;9pii: E1186. doi:10.3390/nu9111186.

7. Tonstad S, Nathan E, Oda K, Fraser G. Vegan diets and hypothyroidism. *Nutrients.* 2013;5:4642-4652.

8. Tonstad S, Nathan E, Oda K, Fraser GE. Prevalence of hyperthyroidism according to type of vegetarian diet. *Public Health Nutr.* 2015;18(8):1482-1487.

9. Messina M, Redmond G. Effects of soy protein and soybean isoflavones on thyroid function in healthy adults and hypothyroid patients: a review of the relevant literature. *Thyroid.* 2006;16(3):249-258.

10. Bitto A, Polito F, Atteritano M, et al. Genistein aglycone does not affect thyroid function: results from a three-year, randomized, double-blind, placebo-controlled trial. *J Clin Endocrinol Metab.* 2010;95(6):3067-3072.

11. Mittal N, Hota D, Dutta P, et al. Evaluation of effect of isoflavone on thyroid economy & autoimmunity in oophorectomised women: a randomised, double-blind, placebo-controlled trial. *Indian J Med Res.* 2011;133(6):633-640.

12. Levis S, Strickman-Stein N, Ganjei-Azar P, Xu P, Doerge DR, Krischer J. Soy isoflavones in the prevention of menopausal bone loss and menopausal symptoms: a randomized, double-blind trial. *Arch Intern Med.* 2011;171(15):1363-1369.

13. Sosvorová L, Mikšátková P, Bičíková M, Kaňová N, Lapčík O. The presence of monoiodinated derivates of daidzein and genistein in human urine and its effect on thyroid gland function. *Food Chem Toxicol.* 2012;50(8):2774-2779.

14. Alekel DL, Genschel U, Koehler KJ, et al. Soy Isoflavones for Reducing Bone Loss Study: effects of a 3-year trial on hormones, adverse events, and endometrial thickness in postmenopausal women. *Menopause.* 2015;22(2):185-197.

15. Tonstad S, Jaceldo-Siegl K, Messina M, Haddad E, Fraser GE. The association between soya consumption and serum thyroid-stimulating hormone concentrations in the Adventist Health Study-2. *Public Health Nutr.* 2016;19(8):1464-1470.

16. Liwanpo L, Hershman JM. Conditions and drugs interfering with thyroxine absorption. *Best Pract Res Clin Endocrinol Metab.* 2009;23(6):781-792.

17. Liel Y, Harman-Boehm I, Shany S. Evidence for a clinically important adverse effect of fiber-enriched diet on the bioavailability of levothyroxine in adult hypothyroid patients. *J Clin Endocrinol Metab.* 1996;81:857-859.

Chapter 10: Healthy Skin and Hair

1. Angela P. Does chocolate cause acne? Verywell Health. https://www.very wellhealth.com/does-chocolate-cause-acne-15519. Accessed November 7, 2018.

2. WebMD. Chocolate and acne. https://www.webmd.com/skin-problems -and-treatments/features/chocolate-and-acne. Accessed November 7, 2018.

3. Fulton JE, Plewig G, Kligman AM. Effect of chocolate on acne vulgaris. *JAMA.* 1969;210(11):2071-2074.

4. Denise G. Dr. Albert M. Kligman, dermatologist, dies at 93. *New York Times.* February 22, 2010. https://www.nytimes.com/2010/02/23/us/23kligman.html.

5. Vongraviopap S, Asawanonda P. Dark chocolate exacerbates acne. *Int J Dermatol.* 2016;55(5):587-591.

6. Block SG, Valins WE, Caperton CV, Viera MH, Amini S, Berman B. Exacerbation of facial acne vulgaris after consuming pure chocolate. *J Am Acad Dermatol.* 2011;65(4):e114-e115.

7. Chalyk N, Klochkov V, Sommereux L, Bandaletova T, Kyle N, Petyaev I. Continuous dark chocolate consumption affects human facial skin surface by stimulating corneocyte desquamation and promoting bacterial colonization. *J Clin Aesthet Dermatol.* 2018;11(9):37-41.

8. Caperton C, Block S, Viera M, Keri J, Berman B. Double-blind, placebo-controlled study assessing the effect of chocolate consumption in subjects with a history of acne vulgaris. *J Clin Aesthet Dermatol.* 2014;7(5):19-23.

9. Delost GR, Delost ME, Lloyd J. The impact of chocolate consumption on acne vulgaris in college students: a randomized crossover study. *J Am Acad Dermatol.* 2016;75(1):220-222.

10. Adebamowo CA, Spiegelman D, Danby FW, Frazier AL, Willett WC, Holmes MD. High school dietary intake and teenage acne. *J Am Acad Dermatol.* 2005;52:207-211.

11. Adebamowo C, Spiegelman D, Berkey CS, et al. Milk consumption and acne in adolescent girls. *Dermatol Online J.* 2006;12(4):1-13.

12. Adebamowo C, Spiegelman D, Berkey CS, et al. Milk consumption and acne in teenaged boys. *J Am Acad Dermatol.* 2008;58(5):787-793.

13. Danby FW. Acne and milk, the diet myth, and beyond. *J Am Acad Dermatol.* 2005;52:360-362.

14. Melnik BC, Schmitz G. Role of insulin, insulin-like growth factor-1, hyperglycaemic food and milk consumption in the pathogenesis of acne vulgaris. *Exp Dermatol.* 2009;18(10):833-841.

15. Cordain L, Lindeberg S, Hurtado M, Hill K, Eaton SB, Brand-Miller J. Acne vulgaris: a disease of Western civilization. *Arch Dermatol.* 2002;138(12):1584-1590.

16. Bendiner E. Disastrous trade-off: Eskimo health for white "civilization." *Hosp Pract.* 1974;9:156-189.

17. Steiner PE. Necropsies on Okinawans. *Arch Pathol.* 1946;42(4):359-380.

18. Smith RN, Mann NJ, Braue A, Mäkeläinen H, Varigos GA. The effect of a high-protein, low glycemic-load diet versus a conventional, high glycemic-load diet on biochemical parameters associated with acne vulgaris: a randomized, investigator-masked, controlled trial. *J Am Acad Dermatol.* 2007;57(2):247-256.

19. Hamilton JB. Male hormone stimulation is prerequisite and an incitant in common baldness. *Am J Anatomy.* 1942;71:451-480.

20. Wang TL, Zhou C, Shen YW, et al. Prevalence of androgenetic alopecia in China: a community-based study in six cities. *Br J Dermatol.* 2010; 162(4):843-847.

21. Inaba M. Can human hair grow again? Tokyo: Azabu Shokan, Inc., 1985.

22. Jang WS, Son IP, Yeo IK, et al. The annual changes of clinical manifestation of androgenetic alopecia clinic in Korean males and females: an outpatient-based study. *Ann Dermatol.* 2013;25(2):181-188.

23. Bakry OA, Shoeib MA, El Shafiee MK, Hassan A. Androgenetic alopecia, metabolic syndrome, and insulin resistance: Is there any association? A case-control study. *Indian Dermatol Online J.* 2014;5(3):276-281.

24. Matilainen V, Laakso M, Hirsso P, Koskela P, Rajala U, Keinänen-Kiukaanniemi S. Hair loss, insulin resistance, and heredity in middle-aged women. A population-based study. *J Cardiovasc Risk.* 2003;10(3):227-231.

Chapter 11: Foods That Fight Moodiness and Stress

1. Beezhold BL, Johnston CS, Daigle DR. Vegetarian diets are associated with healthy mood states: a cross-sectional study in Seventh Day Adventist adults. *Nutr J.* 2010;9:26.

2. Beezhold B, Radnitz C, Rinne A, DiMatteo J. Vegans report less stress and anxiety than omnivores. *Nutr Neurosci.* 2015;18(7):289-296.

3. Sánchez-Villegas A, Delgado-Rodríguez M, Alonso A, et al. Association of the Mediterranean dietary pattern with the incidence of depression: the Seguimiento Universidad de Navarra/University of Navarra follow-up (SUN) cohort. *Arch Gen Psychiatry.* 2009;66(10):1090-1098.

4. Sánchez-Villegas A, Henríquez-Sánchez P, Ruiz-Canela M, et al. A longitudinal analysis of diet quality scores and the risk of incident depression in the SUN Project. *BMC Med.* 2015;13:197.

5. Lassale C, Batty GD, Baghdadli A, et al. Healthy dietary indices and risk of depressive outcomes: a systematic review and meta-analysis of observational studies. *Mol Psychiatry.* 2018 Sep 26. doi:10.1038/s41380-018-0237-8.

6. Tsai AC, Chang TL, Chi SH. Frequent consumption of vegetables predicts lower risk of depression in older Taiwanese—results of a prospective population-based study. *Public Health Nutr.* 2012;15(6):1087-1092.

7. Ocean N, Howley P, Ensor J. Lettuce be happy: a longitudinal UK study on the relationship between fruit and vegetable consumption and well-being. *Soc Sci Med.* 2019;222:335-345.

8. Lindström LH, Nyberg F, Terenius L, et al. CSF and plasma β-casomorphin-like opioid peptides in postpartum psychosis. *Am J Psychiatry.* 1984;141(9):1059-1066.

9. Nyberg F, Lieberman H, Lindström LH, Lyrenäs S, Koch G, Terenius L. Immunoreactive β-casomorphin-8 in cerebrospinal fluid from pregnant and lactating women: correlation with plasma levels. *J Clin Encrinol Metab.* 1989;68:283-289.

10. Beezhold BL, Johnston CS. Restriction of meat, fish, and poultry in omnivores improves mood: a pilot randomized controlled trial. *Nutr J.* 2012;11:9.

11. Ferdowsian HR, Barnard ND, Hoover VJ, et al. A multi-component intervention reduces body weight and cardiovascular risk at a GEICO corporate site. *Am J Health Promot.* 2010;24:384-387.

12. Katcher HI, Ferdowsian HR, Hoover VJ, Cohen JL, Barnard ND. A worksite vegan nutrition program is well-accepted and improves health-related quality of life and work productivity. *Ann Nutr Metab.* 2010;56:245-252.

13. Mishra S, Xu J, Agarwal U, Gonzales J, Levin S, Barnard N. A multicenter randomized controlled trial of a plant-based nutrition program to reduce body weight and cardiovascular risk in the corporate setting: the GEICO study. *Eur J Clin Nutr.* 2013;67:718-724.

14. Agarwal U, Mishra S, Xu J, Levin S, Gonzales J, Barnard N. A multicenter randomized controlled trial of a nutrition intervention program in a multiethnic adult population in the corporate setting reduces depression and anxiety and improves quality of life: the GEICO study. *Am J Health Promot*. 2015;29(4):245-254.

15. Silberman A, Banthia R, Estay IS, et al. The effectiveness and efficacy of an intensive cardiac rehabilitation program in 24 sites. *Am J Health Promot*. 2010;24(4):260-266.

16. Brinkworth GD, Buckley JD, Noakes M, Clifton PM, Wilson CJ. Long-term effects of a very low-carbohydrate diet and a low-fat diet on mood and cognitive function. *Arch Intern Med*. 2009;169(20):1873-1880.

17. Brinkworth GD, Luscombe-Marsh ND, Thompson CH, et al. Long-term effects of very low-carbohydrate and high-carbohydrate weight-loss diets on psychological health in obese adults with type 2 diabetes: randomized controlled trial. *J Intern Med*. 2016;280:388-397.

18. Christopher H. How Japan came to believe in depression. BBC News, July 20, 2016. https://www.bbc.com/news/magazine-36824927. Accessed October 7, 2018.

19. Kjeldsen-Kragh J, Mellbye OJ, Haugen M, et al. Changes in laboratory variables in rheumatoid arthritis patients during a trial of fasting and one-year vegetarian diet. *Scand J Rheumatol*. 1995;24(2):85-93.

20. McDougall J, Bruce B, Spiller G, Westerdahl J, McDougall M. Effects of a very low-fat, vegan diet in subjects with rheumatoid arthritis. *J Altern Complement Med*. 2002;8(1):71-75.

21. Chiavaroli L, Nishi SK, Khan TA, et al. Portfolio dietary pattern and cardiovascular disease: a systematic review and meta-analysis of controlled trials. *Prog Cardiovasc Dis*. 2018;61:43-53.

22. Dowlati Y, Herrmann N, Swardfager W, et al. A meta-analysis of cytokines in major depression. *Biol Psychiatry*. 2010;67(5):446-457.

23. Berk M, Williams LJ, Jacka FN, et al. So depression is an inflammatory disease, but where does the inflammation come from? *BMC Med*. 2013;11.

24. Leonard BE. Inflammation and depression: a causal or coincidental link to the pathophysiology? *Acta Neuropsychiatr*. 2018;30(1):1-16.

25. Farooqui AA, Horrocks LA, Farooqui T. Modulation of inflammation in brain: a matter of fat. *J Neurochem*. 2007;101(3):577-599.

26. Messina M, Gleason C. Evaluation of the potential antidepressant effects of soybean isoflavones. *Menopause*. 2016;23(12):1348-1360.

27. Xu H, Li S, Song X, Li Z, Zhang D. Exploration of the association between dietary fiber intake and depressive symptoms in adults. *Nutrition*. 2018;54:48-53.

28. Liang S, Wu X, Hu X, Wang T, Jin F. Recognizing depression from the microbiota–gut–brain axis. *Int J Mol Sci*. 2018;19:1592. doi:10.3390/ijms19061592.

29. Young SN. Folate and depression—a neglected problem. *J Psychiatry Neurosci*. 2007;32(2):80-82.

30. Coppen A, Bolander-Gouaille C. Treatment of depression: time to consider folic acid and vitamin B12. *J Psychopharmacol.* 2005;19(1):59-65.

31. Blumenthal JA, Babyak MA, Moore KA, et al. Effects of exercise training on older patients with major depression. *Arch Intern Med.* 1999;159:2349-2356.

32. Dunn AL, Trivedi MH, Kampert JB, Clark CG, Chambliss HO. The DOSE study: a clinical trial to examine efficacy and dose response of exercise as treatment for depression. *Control Clin Trials.* 2002;23(5):584-603.

33. Dunn AL, Trivedi MH, Kampert JB, et al. Exercise treatment for depression efficacy and dose response. *Am J Prev Med.* 2005;28(1):1-8.

34. Schuch FB, Vancampfort D, Firth J, et al. Physical activity and incident depression: a meta-analysis of prospective cohort studies. *Am J Psychiatry.* 2018;175(7):631-648.

35. Erickson KI, Voss MW, Prakash RS, et al. Exercise training increases size of hippocampus and improves memory. *Proc Natl Acad Sci USA.* 2011;108: 3017-3022.

Chapter 12: A Healthy Diet

1. Bolland M, Grey A. Clinical trial evidence and use of fish oil supplements. *JAMA Int Med.* 2014;174(3):460-462.

2. Cardoso C, Afonso C, Bandarra NM. Dietary DHA and health: cognitive function ageing. *Nutr Res Rev.* 2016;29(2):281-294.

Chapter 13: Avoiding Environmental Chemicals

1. Carwile JL, Ye X, Zhou X, Calafat AM, Michels KB. Canned soup consumption and urinary bisphenol A: a randomized crossover trial. *JAMA.* 2011;306(20):2218-2220.

2. Lang IA, Galloway TS, Scarlett A, et al. Association of urinary bisphenol A concentration with medical disorders and laboratory abnormalities in adults. *JAMA.* 2008;300(11):1303-1310.

3. Ehrlich S, Calafat AM, Humblet O, Smith T, Hauser R. Handling of thermal receipts as a source of exposure to bisphenol A. *JAMA.* 2014;311(8):859-860.

4. Li DK, Zhou Z, Miao M, et al. Urine bisphenol-A (BPA) level in relation to semen quality. *Fertil Steril.* 2011;95(2):625-630.

5. Li DK, Zhou Z, Miao M, et al. Relationship between urine bisphenol-A level and declining male sexual function. *J Androl.* 2010;31(5):500-506.

6. Dziewirska E, Hanke W, Jurewicz J. Environmental non-persistent endocrine-disrupting chemicals exposure and reproductive hormones levels in adult men. *Int J Occup Med Environ Health.* 2018;31(5). doi.org/10.13075/ijomeh.1896.01183.

7. Pollack AZ, Mumford SL, Krall JR, et al. Exposure to bisphenol A, chlorophenols, benzophenones, and parabens in relation to reproductive hormones in healthy women: a chemical mixture approach. *Environ Int.* 2018;120:137-144.

8. Hu Y, Wen S, Yuan D, et al. The association between the environmental endocrine disruptor bisphenol A and polycystic ovary syndrome: a systematic review and meta-analysis. *Gynecol Endocrinol.* 2018;34(5):370-377.

9. Trasande L, Attina TM, Blustein J. Association between urinary bisphenol A concentration and obesity prevalence in children and adolescents. *JAMA.* 2012;308(11):1113-1121.

10. Luo Q, Liu ZH, Yin H, et al. Migration and potential risk of trace phthalates in bottled water: a global situation. *Water Res.* 2018;147:362-372.

11. Erythropel HC, Maric M, Nicell JA, Leask RL, Yargeau V. Leaching of the plasticizer di(2-ethylhexyl)phthalate (DEHP) from plastic containers and the question of human exposure. *Appl Microbiol Biotechnol.* 2014;98(24):9967-9981.

12. Zota AR, Phillips CA, Mitro SD. Recent fast food consumption and bisphenol A and phthalates exposures among the U.S. population in NHANES, 2003–2010. *Environ Health Perspect.* 2016;124(10):1521-1528.

13. Braun JM, Sathyanarayana S, Hauser R. Phthalate exposure and children's health. *Curr Opin Pediatr.* 2013;25(2):247-254.

14. Ejaredar M, Nyanza EC, Ten Eycke K, Dewey D. Phthalate exposure and children's neurodevelopment: a systematic review. *Environ Res.* 2015;142:51-60.

15. Attina TM, Trasande L. Association of exposure to di-2-ethylhexylphthalate replacements with increased insulin resistance in adolescents from NHANES 2009–2012. *J Clin Endocrinol Metab.* 2015;100(7):2640-2650.

16. Trasande L, Attina TM. Association of exposure to di-2-ethylhexylphthalate replacements with increased blood pressure in children and adolescents. *Hypertension.* 2015;66(2):301-308.

17. James-Todd T, Stahlhut R, Meeker JD, et al. Urinary phthalate metabolite concentrations and diabetes among women in the National Health and Nutrition Examination Survey (NHANES) 2001–2008. *Environ Health Perspect.* 2012;120(9):1307-1313.

18. James-Todd TM, Huang T, Seely EW, Saxena AR. The association between phthalates and metabolic syndrome: the National Health and Nutrition Examination Survey 2001–2010. *Environ Health.* 2016;15:52.

19. Thongprakaisang S, Thiantanawat A, Rangkadilok N, Suriyo T, Satayavivad J. Glyphosate induces human breast cancer cells growth via estrogen receptors. *Food Chem Toxicol.* 2013;59:129-136.

20. International Agency for Research on Cancer, World Health Organization. IARC Monographs Volume 112: evaluation of five organophosphate insecticides and herbicides. 20 March 2015. https://www.iarc.fr/en/media-centre/iarcnews/pdf/MonographVolume112.pdf.

21. European Food Safety Authority. Glyphosate: EFSA updates toxicological profile. 12 November 2015. https://www.efsa.europa.eu/en/press/news/151112.

22. Almberg KS, Turyk ME, Jones RM, Rankin K, Freels S, Stayner LT. Atrazine contamination of drinking water and adverse birth outcomes in community

water systems with elevated atrazine in Ohio, 2006–2008. *Int J Environ Res Public Health*. 2018;15(9). pii: E1889. doi: 10.3390/ijerph15091889.

23. Orton F, Rosivatz E, Scholze M, Kortenkamp A. Widely used pesticides with previously unknown endocrine activity revealed as *in vitro* antiandrogens. *Environ Health Perspect*. 2011;119(6). doi.org/10.1289/ehp.1002895.

24. Bretveld RW, Thomas CM, Scheepers PT, Zielhuis GW, Roeleveld N. Pesticide exposure: the hormonal function of the female reproductive system disrupted? *Reprod Biol Endocrinol*. 2006;4:30. doi:10.1186/1477-7827-4-30.

25. US Environmental Protection Agency. Learn about Polychlorinated Biphenyls (PCBs). https://www.epa.gov/pcbs/learn-about-polychlorinated-biphenyls-pcbs. Accessed September 11, 2017.

26. Jacobson JL, Jacobson SW. Intellectual impairment in children exposed to polychlorinated biphenyls in utero. *N Engl J Med*. 1996;335:783-789.

27. Agency for Toxic Substances & Disease Registry. Toxicological profile for polychlorinated biphenyls (PCBs). http://www.atsdr.cdc.gov/ToxProfiles/TP.asp?id=142&tid=26. Accessed August 1, 2016.

28. World Health Organization. Dioxins and their effects on human health. http://www.who.int/mediacentre/factsheets/fs225/en/index.html. Accessed September 11, 2017.

29. Sweis IE, Cressey BC. Potential role of the common food additive manufactured citric acid in eliciting significant inflammatory reactions contributing to serious disease states: a series of four case reports. *Toxicol Rep*. 2018;5:808-812.

30. Rivera-Núñez Z, Barrett ES, Szamreta EA, et al. Urinary mycoestrogens and age and height at menarche in New Jersey girls. *Environ Health*. 2019;18(1):24. doi: 10.1186/s12940-019-0464-8.

31. Hergenrather J, Hlady G, Wallace B, Savage E. Pollutants in breast milk of vegetarians. *Lancet*. 1981;304:792.

32. Kahleova H, Tonstad S, Rosmus J, et al. The effect of a vegetarian versus conventional hypocaloric diet on serum concentrations of persistent organic pollutants in patients with type 2 diabetes. *Nutr Metab Cardiovasc Dis*. 2016;26(5):430-438.

33. Environmental Working Group. *EWG's 2018 Shopper's Guide to Pesticides in Produce*. https://www.ewg.org/foodnews/dirty-dozen.php.

34. Baudry J, Assmann KE, Touvier M, et al. Association of frequency of organic food consumption with cancer risk: findings from the NutriNet-Santé Prospective Cohort Study. *JAMA Intern Med*. 2018. doi:10.1001/jamainternmed.2018.4357.

35. Key S, Ma JKC, Drake PMW. Genetically modified plants and human health. *J R Soc Med*. 2008;101:290-298.

36. Magaña-Gómez JA, Calderón de la Barca AM. Risk assessment of genetically modified crops for nutrition and health. *Nutr Rev*. 2009;67:1-16.

Index

About the Author

Neal D. Barnard, MD, FACC, is perhaps the world's most respected authority on vegan diets. He is a faculty member of the George Washington University School of Medicine and president of the Physicians Committee for Responsible Medicine. Dr. Barnard's research, funded by the U.S. government, revolutionized the dietary approach to type 2 diabetes, and his findings have influenced the policies of the U.S. Dietary Guidelines Advisory Committee, the American Diabetes Association, and the American Medical Association. In 2015, he was named a fellow of the American College of Cardiology, and in 2016, he received the American College of Lifestyle Medicine's Trailblazer Award. Dr. Barnard is editor-in-chief of the *Nutrition Guide for Clinicians*, a nutrition textbook given to all medical students in the United States.